THERMODYNAMICS AND KINETICS FOR THE BIOLOGICAL SCIENCES

THERMODYNAMICS AND KINETICS FOR THE BIOLOGICAL SCIENCES

Gordon G. Hammes

Department of Biochemistry
Duke University

A JOHN WILEY & SONS, INC., PUBLICATION

New York • Chichester • Weinheim • Brisbane • Singapore • Toronto

Copyright © 2000 by John Wiley & Sons, Inc. All rights reserved.

Published simultaneously in Canada.

For ordering and customer service, call 1-800-CALL-WILEY.

Library of Congress Cataloging-in-Publication Data:

Hammes, Gordon., 1934–
    Thermodynamics and kinetics for the biological sciences/by Gordon G. Hammes.
        p.   cm.
    "Published simultaneously in Canada."
    Includes bibliographical references and index.
    ISBN 0-471-37491-1 (pbk.: acid-free paper)
    1. Physical biochemistry. 2. Thermodynamics. 3. Chemical kinetics. I. Title.
    QP517.P49 H35 2000
    572—dc21   99-086233

Printed in the United States of America.

10   9   8   7   6   5

CONTENTS

Appendixes

This book is based on a course that I have been teaching for the past several years to first year graduate students in the biological sciences at Duke University. These students have not studied physical chemistry as undergraduates and typically have not had more than a year of calculus. Many faculty believe that an understanding of the principles of physical chemistry is important for all students in the biological sciences, and this course is required by the Cell and Molecular Biology Program. The course consists of two parts—one devoted to thermodynamics and kinetics, the other to spectroscopy. Only the first half of the course is covered in this volume. An introduction to spectroscopy is being planned as a separate volume. One of the reviewers of the proposal for this book said that it was impossible to teach biology students this material—the reviewer had been trying for many years. On the contrary, I believe the students that have taken this course have mastered the principles of the subject matter and will find the knowledge useful in their research.

Thermodynamics and kinetics are introduced with a minimum of mathematics. However, the approach is quantitative and is designed to introduce the student to the important concepts that are necessary to apply the principles of thermodynamics and kinetics to biology. The applications cover a wide range of topics and vary considerably in the degree of difficulty. More material is included than is covered in the course on which the book is based, which will allow the students and instructors to pick and choose. Some problems are also included, as problem solving is an important part of understanding principles.

I am indebted to my colleagues at Duke University for their encouragement and assistance. Discussions with them were essential to the production of this book. Special thanks are due to Professor Jane Richardson and Dr. Michael Word for their assistance with the color figures. I also want to acknowledge the encouragement and assistance of my wife, Judy, during this entire project. I would appreciate any comments or suggestions from the readers of this volume.

GORDON G. HAMMES
Duke University
Durham, North Carolina

CHAPTER 1

Heat, Work, and Energy

1.1 INTRODUCTION

Thermodynamics is deceptively simple or exceedingly complex, depending on how you approach it. In this book, we will be concerned with the principles of thermodynamics that are especially useful in thinking about biological phenomena. The emphasis will be on concepts, with a minimum of mathematics. Perhaps an accurate description might be rigor without *rigor mortis*. This may cause some squirming in the graves of thermodynamic purists, but the objective is to provide a foundation for researchers in experimental biology to use thermodynamics. This includes cell biology, microbiology, molecular biology, and pharmacology, among others. In an ideal world, researchers in these fields would have studied a year of physical chemistry, and this book would be superfluous. Although most biochemists have this background, it is unusual for other biological sciences to require it. Excellent texts are available that present a more advanced and complete exposition of thermodynamics (cf. Refs. 1 and 2).

In point of fact, thermodynamics can provide a useful way of thinking about biological processes and is indispensable when considering molecular and cellular mechanisms. For example, what reactions and coupled physiological processes are possible? What are the allowed mechanisms involved in cell division, in protein synthesis? What are the thermodynamic considerations that cause proteins, nucleic acids, and membranes to assume their active structures? It is easy to postulate biological mechanisms that are inconsistent with thermodynamic principles—but just as easy to postulate those that are consistent. Consequently, no active researcher in biology should be without a rudimentary knowledge of the principles of thermodynamics. The ultimate goal of this exposition is to understand what determines equilibrium in biological systems, and how these equilibrium processes can be coupled together to produce living systems, even though we recognize that living organisms are not at equilibrium. Thermodynamics provides a unifying framework for diverse systems in biology. Both a qualitative and quantitative understanding are important and will be developed.

The beauty of thermodynamics is that a relatively small number of postulates can be used to develop the entire subject. Perhaps the most important part of this development is to be very precise with regard to concepts and definitions, without getting bogged down with mathematics. Thermodynamics is a macroscopic theory, not molecular. As far as thermodynamics is concerned, molecules need not exist. However, we will not be purists in this regard: If molecular descriptions are useful for understanding or introducing concepts, they will be used. We will not hesitate to give molecular descriptions of thermodynamic results, but we should recognize that these interpretations are not inherent in thermodynamics itself. It is important to note, nevertheless, that large collections of molecules are assumed so that their behavior is governed by Boltzmann statistics; that is, the normal thermal energy distribution is assumed. This is almost always the case in practice. Furthermore, thermodynamics is concerned with time-independent systems, that is, systems at equilibrium. Thermodynamics has been extended to nonequilibrium systems, but we will not be concerned with the formal development of this subject here.

The first step is to define the *system*. A thermodynamic system is simply that part of the universe in which we are interested. The only caveat is that the system must be large relative to molecular dimensions. The system could be a room, it could be a beaker, it could be a cell, etc. An *open system* can exchange energy and matter across its boundaries, for example, a cell or a room with open doors and windows. A *closed system* can exchange energy but not matter, for example, a closed room or box. An *isolated system* can exchange neither energy nor matter, for example, the universe or, approximately, a closed Dewar. We are free to select the system as we choose, but it is very important that we specify what it is. This will be illustrated as we proceed. The *properties* of a system are any measurable quantities characterizing the system. Properties are either *extensive*, proportional to the mass of the system, or *intensive*, independent of the mass. Examples of extensive properties are mass and volume. Examples of intensive properties are temperature, pressure, and color.

1.2 TEMPERATURE

We are now ready to introduce three important concepts: temperature, heat, and work. None of these are unfamiliar, but we must define them carefully so that they can be used as we develop thermodynamics.

Temperature is an obvious concept, as it simply measures how hot or cold a system is. We will not belabor its definition and will simply assert that thermodynamics requires a unique temperature scale, namely, the Kelvin temperature scale. The Kelvin temperature scale is related to the more conventional Celsius temperature scale by the definition

$$T_{\text{Kelvin}} = T_{\text{Celsius}} + 273.16 \qquad (1\text{-}1)$$

Although the temperature on the Celsius scale is referred to as "degrees Celsius," by convention degrees are not stated on the Kelvin scale. For example, a temperature of 100 degrees Celsius is 373 Kelvin. (Thermodynamics is entirely logical—some of the conventions used are not.) The definition of *thermal equilibrium* is very simple: when two systems are at the same temperature, they are at thermal equilibrium.

1.3 HEAT

Heat flows across the system boundary during a change in the state of the system because a temperature difference exists between the system and its surroundings. We know of many examples of heat: Some chemical reactions produce heat, such as the combustion of gas and coal. Reactions in cells can produce heat. By convention, heat flows from higher temperature to lower temperature. This fixes the sign of the heat change. It is important to note that this is a convention and is not required by any principle. For example, if the temperature of the surroundings decreases, heat flows to the system, and the sign of the heat change is positive (+). A simple example will illustrate this sign convention as well as the importance of defining the system under consideration.

Consider two beakers of the same size filled with the same amount of water. In one beaker, A, the temperature is 25°C, and in the other beaker, B, the temperature is 75°C. Let us now place the two beakers in thermal contact and allow them to reach thermal equilibrium (50°C). This situation is illustrated in Figure 1-1. If the system is defined as A, the temperature of the system increases so the heat change is positive. If the system is defined as B, the temperature of the system decreases so the heat change is negative. If the system is defined as A and B, no heat flow occurs across the boundary of the system, so the heat change is zero! This illustrates how important it is to define the system before asking questions about what is occurring.

The heat change that occurs is proportional to the temperature difference between the initial and final states of the system. This can be expressed mathematically as

$$q = C(T_f - T_i) \tag{1-2}$$

where q is the heat change, the constant C is the *heat capacity*, T_f is the final temperature, and T_i is the initial temperature. This relationship assumes that the heat capacity is constant, independent of the temperature. In point of fact, the heat capacity often changes as the temperature changes, so that a more precise definition puts this relationship in differential form:

$$dq = C \, dT \tag{1-3}$$

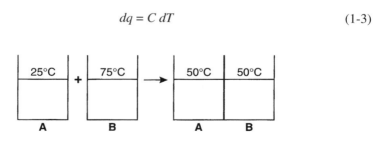

FIGURE 1-1. Illustration of the establishment of thermal equilibrium and importance of defining the *system* carefully. Two identical vessels filled with the same amount of liquid, but at different temperatures, are placed in contact and allowed to reach thermal equilibrium. A discussion of this figure is given in the text.

Note that the heat change and the heat capacity are extensive properties—the larger the system the larger the heat capacity and the heat change. Temperature, of course, is an intensive property.

1.4   WORK

The definition of *work* is not as simple as that for heat. Many different forms of work exist, for example, mechanical work, such as muscle action, and electrical work, such as ions crossing charged membranes. We will use a rather artificial, but very general, definition of work that is easily understood. Work is a quantity that can be transferred across the system boundary and can always be converted to lifting and lowering a weight in the surroundings. By convention, work done on a system is positive: this corresponds to lowering the weight in the surroundings.

You may recall that mechanical work, w, is defined as the product of the force in the direction of movement, F_x, times the distance moved, x, or in differential form

$$dw = F_x \, dx \tag{1-4}$$

Therefore, the work to lower a weight is $-mgh$, where m is the mass, g is the gravitational constant, and h is the distance the weight is lowered. This formula is generally useful: For example, mgh is the work required for a person of mass m to walk up a hill of height h. The work required to stretch a muscle could be calculated with Eq. 1-4 if we knew the force required and the distance the muscle was stretched. Electrical work, for example, is equal to $-EIt$, where E is the electromotive force, I is the current, and t is the time. In living systems, membranes often have potentials (voltages) across them. In this case, the work required for an ion to cross the membrane is $-zF\Psi$, where z is the valence of the ion, F is the Faraday (96,489 coulombs per mole), and Ψ is the potential. A specific example is the cotransport of Na^+ and K^+, Na^+ moving out of the cell and K^+ moving into the cell. A potential of -70 millivolts is established on the inside so that the electrical work required to move a mole of K^+ ions to the inside is $-(1)(96,489)(0.07) = -6750$ joules. ($\Psi = \Psi_{outside} - \Psi_{inside} = +70$ millivolts.) The negative sign means that work is done by the system.

Although not very biologically relevant, we will now consider in some detail pressure–volume work, or P–V work. This type of work is conceptually easy to understand, and calculations are relatively easy. The principles discussed are generally applicable to more complex systems, such as those encountered in biology. As a simple example of P–V work, consider a piston filled with a gas, as pictured in Figure 1-2. In this case, the force is equal to the external pressure, P_{ex}, times the area, A, of the piston face, so the infinitesimal work can be written as

$$dw = -P_{ex} A \, dx = -P_{ex} \, dV \tag{1-5}$$

If the piston is lowered, work is done on the system and is positive; whereas if the piston is raised work is done by the system and is negative. Note that the work done on

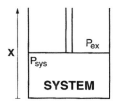

FIGURE 1-2. Schematic representation of a piston pushing on the system. P_{ex} is the external pressure, and P_{sys} is the pressure of the system.

or by the system by lowering or raising the piston depends on what the external pressure is. Therefore, the work can have any value from 0 to ∞, depending on how the process is done. This is a very important point: The work associated with a given change in state depends on *how* the change in state is carried out.

The idea that work depends on how the process is carried out can be illustrated further by considering the expansion and compression of a gas. The $P-V$ isotherm for an ideal gas is shown in Figure 1-3. An ideal gas is a gas that obeys the ideal gas law, $PV = nRT$ (n is the number of moles of gas and R is the gas constant). The behavior of most gases at moderate pressures is well described by this relationship. Let us consider the expansion of the gas from P_1, V_1 to P_2, V_2. If this expansion is done with the external pressure equal to zero, that is, into a vacuum, the work is zero. Clearly this is the minimum amount of work that can be done for this change in state. Let us now carry out the same expansion with the external pressure equal to P_2. In this case, the work is

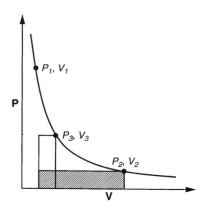

FIGURE 1-3. A $P-V$ isotherm for an ideal gas. The narrow rectangle with both hatched and open areas is the work done in going from P_1,V_1 to P_3,V_3 with an external pressure of P_3. The hatched area is the work done by the system in going from P_1,V_1 to P_2,V_2 with an external pressure of P_2. The maximum amount of work done by the system for this change in state is the area under the curve between P_1,V_1 and P_2,V_2.

$$w = -\int_{V_1}^{V_2} P_{ex}\, dV = -P_2(V_2 - V_1) \qquad (1\text{-}6)$$

which is the striped area under the P–V curve. The expansion can be broken into stages; for example, first expand the gas with $P_{ex} = P_3$ followed by $P_{ex} = P_2$, as shown in Figure 1-3. The work done by the system is then the sum of the two rectangular areas under the curve. It is clear that as the number of stages is increased, the magnitude of the work done increases. The maximum work that can be attained would set the external pressure equal to the pressure of the system minus a small differential pressure, dP, throughout the expansion. This can be expressed as

$$w_{max} = -\int_{V_1}^{V_2} P\, dV \qquad (1\text{-}7)$$

By a similar reasoning process, it can be shown that for a compression the minimum work done on the system is

$$w_{min} = -\int_{V_2}^{V_1} P\, dV \qquad (1\text{-}8)$$

This exercise illustrates two important points. First, it clearly shows that the work associated with a change in state depends on how the change in state is carried out. Second, it demonstrates the concept of a *reversible path*. When a change in state is carried out such that the surroundings and the system are not at equilibrium by an infinitesimal amount, in this case dP, during the change in state, the process is called reversible. The concept of reversibility is only an ideal—it cannot be achieved in practice. Obviously we cannot really carry out a change in state with only an infinitesimal difference between the pressures of the system and surroundings. We will find this concept very useful, nevertheless.

Now let's think about a cycle whereby an expansion is carried out followed by a compression that returns the system back to its original state. If this is done as a one-stage process in each case, the total work can be written as

$$w_{total} = w_{exp} + w_{comp} \qquad (1\text{-}9)$$

or

$$w_{total} = -P_2(V_2 - V_1) - P_1(V_1 - V_2) \qquad (1\text{-}10)$$

or

$$w_{total} = (P_1 - P_2)(V_2 - V_1) > 0 \qquad (1\text{-}11)$$

In this case, net work has been done on the system. For a reversible process, however, the work associated with compression and expansion is

$$w_{exp} = -\int_{V_1}^{V_2} P \, dV \tag{1-12}$$

and

$$w_{comp} = -\int_{V_2}^{V_1} P \, dV \tag{1-13}$$

so that the total work for the cycle is equal to zero. Indeed, for reversible cycles the net work is always zero.

To summarize this discussion of the concept of work, the work done on or by the system depends on how the change in state of the system occurs. In the real world, changes in state always occur irreversibly, but we will find the concept of a reversible change in state to be very useful.

Heat changes also depend on how the process is carried out. Generally a subscript is appended to q, for example, q_P and q_V for heat changes at constant pressure and volume, respectively. As a case in point, the heat change at constant pressure is greater than that at constant volume if the temperature of a gas is raised. This is because not only must the temperature be raised, but the gas must also be expanded.

Although this discussion of gases seems far removed from biology, the concepts and conclusions reached are quite general and can be applied to biological systems. The only difference is that exact calculations are usually more difficult. It is useful to consider why this is true. In the case of ideal gases, a simple equation of state is known, $PV = nRT$, that is obeyed quite well by real gases under normal conditions. This equation is valid because gas molecules, on average, are quite far apart and their energetic interactions can be neglected. Collisions between gas molecules can be approximated as billiard balls colliding. This situation obviously does not prevail in liquids and solids where molecules are close together and the energetics of their interactions cannot be neglected. Consequently, simple equations of state do not exist for liquids and solids.

1.5   DEFINITION OF ENERGY

The first law of thermodynamics is basically a definition of the energy change associated with a change in state. It is based on the experimental observation that heat and work can be interconverted. Probably the most elegant demonstration of this is the experimental work of James Prescott Joule in the late 1800s. He carried out experiments in which he measured the work necessary to turn a paddle wheel in water and the concomitant rise in temperature of the water. With this rather primitive experiment, he was able to calculate the conversion factor between work and heat with amazing accuracy, namely, to within 0.2%. The first law states that the energy change, ΔE, associated with a change in state is

$$\Delta E = q + w \qquad (1\text{-}14)$$

Furthermore, the energy change is the same regardless of how the change in state is carried out. In this regard, energy clearly has quite different properties than heat and work. This is true for both reversible and irreversible processes. Because of this property, the energy (usually designated the internal energy in physical chemistry textbooks) is called a *state function*. State functions are extremely important in thermodynamics, both conceptually and practically.

Obviously we cannot prove the first law, as it is a basic postulate of thermodynamics. However, we can show that without this law events could occur that are contrary to our experience. Assume, for example, that the energy change in going from state 1 to state 2 is greater than the negative of that for going from state 2 to 1 because the changes in state are carried out differently. We could then cycle between these two states and produce energy as each cycle is completed, essentially making a perpetual motion machine. We know that such machines do not exist, consistent with the first law. Another way of looking at this law is as a statement of the conservation of energy.

It is important that thermodynamic variables are not just hypothetical—we must be able to relate them to laboratory experience, that is, to measure them. Thermodynamics is developed here for practical usage. Therefore, we must be able to relate the concepts to what can be done in the laboratory. How can we measure energy changes? If we only consider P–V work, the first law can be written as

$$\Delta E = q - \int_{V_1}^{V_2} P_{ex}\, dV \qquad (1\text{-}15)$$

If the change in state is measured at constant volume, then

$$\Delta E = q_V \qquad (1\text{-}16)$$

At first glance, it may seem paradoxical that a state function, the energy change, is equal to a quantity whose magnitude depends on how the change in state is carried out, namely, the heat change. However, in this instance we have specified how the change in state is to occur, namely, at constant volume. Therefore, if we measure the heat change at constant volume associated with a change in state, we have also measured the energy change.

Temperature is an especially important variable in biological systems. If the temperature is constant during a change in state, the process is *isothermal*. On the other hand, if the system is insulated so that no heat escapes or enters the system during the change in state ($q = 0$), the process is *adiabatic*.

1.6 ENTHALPY

Most experiments in the laboratory and in biological systems are done at constant pressure, rather than at constant volume. At constant pressure,

$$\Delta E = q_P - P(V_2 - V_1) \tag{1-17}$$

or

$$E_2 - E_1 = q_P - P(V_2 - V_1) \tag{1-18}$$

The heat change at constant pressure can be written as

$$q_P = (E_2 + PV_2) - (E_1 + PV_1) \tag{1-19}$$

This relationship can be simplified by defining a new state function, the *enthalpy*, H:

$$H = E + PV \tag{1-20}$$

The enthalpy is obviously a state function since E, P, and V are state functions. The heat change at constant pressure is then equal to the enthalpy change:

$$q_P = \Delta H = H_2 - H_1 \tag{1-21}$$

For biological reactions and processes, we will usually be interested in the enthalpy change rather than the energy change. It can be measured experimentally by determining the heat change at constant pressure.

As a simple example of how energy and enthalpy can be calculated, let's consider the conversion of liquid water to steam at 100°C and 1 atmosphere pressure, that is, boiling water:

$$H_2O(\ell, 1\text{ atm}, 100°C) \rightarrow H_2O(g, 1\text{ atm}, 100°C) \tag{1-22}$$

The heat required for this process, $\Delta H\ (= q_P)$ is 9.71 kilocalories/mole. What is ΔE for this process? This can be calculated as follows:

$$\Delta E = \Delta H - \Delta(PV) = \Delta H - P\,\Delta V$$

$$\Delta V = V_g - V_\ell = 22.4\text{ liters/mole} - 18.0 \times 10^{-3}\text{ liters/mole} \approx PV_g \approx RT$$

$$\Delta E = \Delta H - RT = 9710 - 2(373) = 8970\text{ calories/mole}$$

Note that the Kelvin temperature must be used in thermodynamic calculations and that ΔH is significantly greater than ΔE.

Let's do a similar calculation for the melting of ice into liquid water

$$H_2O(s, 273 \text{ K}, 1 \text{ atm}) \rightarrow H_2O(\ell, 273 \text{ K}, 1 \text{ atm}) \qquad (1\text{-}23)$$

In this case the measured heat change, ΔH $(= q_p)$ is 1.44 kilocalories/mole. The calculation of ΔE parallels the previous calculation.

$$\Delta E = \Delta H - P\,\Delta V$$

$$\Delta V = V_\ell - V_s \approx 18.0 \text{ milliliters/mole} - 19.6 \text{ milliliters/mole} \approx -1.6 \text{ milliliters/mole}$$

$$P\,\Delta V \approx -1.6 \text{ mL·atm} = -0.04 \text{ calorie}$$

$$\Delta E = 1440 + 0.04 = 1440 \text{ calories/mole}$$

In this case ΔE and ΔH are essentially the same. In general, they do not differ greatly in condensed media, but the differences can be substantial in the gas phase.

The two most common units for energy are the calorie and the joule. (One calorie equals 4.184 joules.) The official MKS unit is the joule, but many research publications use the calorie. We will use both in this text, in order to familiarize the student with both units.

1.7 STANDARD STATES

Only changes in energy states can be measured. Therefore, it is arbitrary what we set as the zero for the energy scale. As a matter of convenience, a common zero has been set for both the energy and enthalpy. Elements in their stablest forms at 25°C (298 K) and 1 atmosphere are assigned an enthalpy of zero. This is called a *standard state* and is usually written as H°_{298}. The superscript means 1 atmosphere, and the subscript is the temperature in Kelvin.

As an example of how this concept is used, consider the formation of carbon tetrachloride from its elements:

$$C \text{ (graphite)} + 2\,Cl_2(g) \rightarrow CCl_4\,(\ell) \qquad (1\text{-}24)$$

$$\Delta H = H^\circ_{298(CCl_4)} - H^\circ_{298(C)} - 2H^\circ_{298(Cl_2)}$$

$$\Delta H = H^\circ_{298(CCl_4)}$$

The quantity $H^\circ_{298(CCl_4)}$ is called the heat of formation of carbon tetrachloride. Tables of heats of formation are available for hundreds of compounds and are useful in calculating the enthalpy changes associated with chemical reactions (cf. 3,4).

In the case of substances of biological interest in solutions, the definitions of standard states and heats of formation are a bit more complex. In addition to pressure and temperature, other factors must be considered such as pH, salt concentration, metal ion concentration, etc. A universal definition has not been established. In practice, it is best to use heats of formation under a defined set of conditions, and likewise to define the standard state as these conditions. Tables of heats of formation for some compounds of biological interest are given in Appendix 1 (3). A prime is often added to the symbol for these heats of formation ($H^{\circ\prime}_f$) to indicate the unusual nature of the standard state. We will not make that distinction here, but it is essential that a consistent standard state is used when making thermodynamic calculations for biological systems.

A useful way of looking at chemical reactions is as algebraic equations. A characteristic enthalpy can be assigned to each product and reactant. Consider the "reaction"

$$a A + b B \rightleftharpoons c C + d D \qquad (1\text{-}25)$$

For this reaction, $\Delta H = H_{\text{products}} - H_{\text{reactants}}$, or

$$\Delta H = d H_D + c H_C - a H_A - b H_B$$

where the H_i are molar enthalpies. At 298 K and 1 atmosphere, the molar enthalpies of the elements are zero, whereas for compounds, the molar enthalpies are equal to the heats of formation, which are tabulated. Before we apply these considerations to biological reactions, a brief digression will be made to discuss how heats of reactions are determined experimentally.

1.8 CALORIMETRY

The area of science concerned with the measurement of heat changes associated with chemical reactions is designated as calorimetry. Only a brief introduction is given here, but it is important to relate the theoretical concepts to laboratory experiments. To begin this discussion we will return to our earlier discussion of heat changes and the heat capacity, Eq. 1-3. Since the heat change depends on how the change in state is carried out, we must be more precise in defining the heat capacity. The two most common conditions are constant volume and constant pressure. The heat changes in these cases can be written as

$$dq_V = dE = C_V \, dT \qquad (1\text{-}26)$$

$$dq_P = dH = C_P \, dT \tag{1-27}$$

A more exact mathematical treatment of these definitions would make use of partial derivatives, but we will avoid this complexity by using subscripts to indicate what is held constant. These equations can be integrated to give

$$\Delta E = \int_{T_1}^{T_2} C_V \, dT \tag{1-28}$$

$$\Delta H = \int_{T_1}^{T_2} C_P \, dT \tag{1-29}$$

Thus, heat changes can readily be measured if the heat capacity is known. The heat capacity of a substance can be determined by adding a known amount of heat to the substance and determining the resulting increase in temperature. The known amount of heat is usually added electrically since this permits very precise measurement. (Recall that the electrical heat is I^2R, where I is the current and R is the resistance of the heating element.) If heat is added repeatedly in small increments over a large temperature range, the temperature dependence of the heat capacity can be determined. Tabulations of heat capacities are available and are usually presented with the temperature dependence described as a power series:

$$C_P = a + bT + cT^2 + \cdots \tag{1-30}$$

where $a, b, c, \ldots$ are constants determined by experiment.

For biological systems, two types of calorimetry are commonly done—batch calorimetry and scanning calorimetry. In batch calorimetry, the reactants are mixed together and the ensuing temperature rise (or decrease) is measured. A simple experimental setup is depicted in Figure 1-4, where the calorimeter is a Dewar flask and the temperature increase is measured by a thermocouple or thermometer.

For example, if we wished to measure the heat change for the hydrolysis of adenosine 5'-triphosphate (ATP),

$$\text{ATP} + \text{H}_2\text{O} \rightleftharpoons \text{ADP} + \text{P}_i \tag{1-31}$$

a solution of known ATP concentration would be put in the Dewar at a defined pH, metal ion concentration, buffer, etc. The reaction would be initiated by adding a small amount of adenosine triphosphatase (ATPase), an enzyme that efficiently catalyzes the hydrolysis, and the subsequent temperature rise measured. The enthalpy of reaction can be calculated from the relationship

$$\Delta H = C_P \, \Delta T \tag{1-32}$$

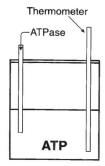

FIGURE 1-4. Schematic representation of a simple batch calorimeter. The insulated vessel is filled with a solution of ATP in a buffer containing salt and Mg^{2+}. The hydrolysis of ATP is initiated by the addition of the ATPase enzyme, and the subsequent rise in temperature is measured.

The heat capacity of the system is calculated by putting a known amount of heat into the system through an electrical heater and measuring the temperature rise of the system. The enthalpy change calculated is for the number of moles of ATP in the system. Usually the experimental result is reported as a molar enthalpy, that is, the enthalpy change for a mole of ATP being hydrolyzed. This result can be obtained by dividing the observed enthalpy change by the moles of ATP hydrolyzed. Actual calorimeters are much more sophisticated than this primitive experimental setup. The calorimeter is well insulated, mixing is done very carefully, and very precise temperature measurements are made with a thermocouple. The enthalpy changes for many biological reactions have been measured, but unfortunately this information is not conveniently tabulated in a single source. However, many enthalpies of reaction can be derived from the heats of formation in the table in Appendix 1.

Scanning calorimetry is a quite different experiment and measures the heat capacity as a function of temperature. In these experiments, a known amount of heat is added to the system through electrical heating and the resulting temperature rise is measured. Very small amounts of heat are used so the temperature changes are typically very small. This process is repeated automatically so that the temperature of the system slowly rises. The heat capacity of the system is calculated for each heat increment as $q_P/\Delta T$, and the data are presented as a plot of C_P versus T.

This method has been used, for example, to study protein unfolding and denaturation. Proteins unfold as the temperature is raised, and denaturation usually occurs over a very narrow temperature range. This is illustrated schematically in Figure 1-5, where the fraction of denatured protein, f_D, is plotted versus the temperature along with the corresponding plot of heat capacity, C_P, versus temperature.

As shown Figure 1-5, the plot of heat capacity versus temperature is a smooth, slowly rising curve for the solvent. With the protein present, a peak in the curve occurs as the protein is denatured. The enthalpy change associated with denaturation is the area under the peak (striped area = $\int C_P \, dT$). In some cases, the protein denaturation

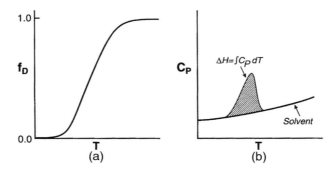

FIGURE 1-5. Schematic representation of the denaturation of a protein and the resulting change in heat capacity, C_P. In (a) the fraction of denatured protein, f_D, is shown as a function of temperature, T. In (b) the heat capacity, as measured by scanning calorimetry, is shown as a function of temperature. The lower curve is the heat capacity of the solvent. The hatched area is the excess heat capacity change due to the protein denaturing and is equal to ΔH for the unfolding.

may occur in multiple stages, in which case more than one peak can be seen in the heat capacity plot. This is shown schematically in Figure 1-6 for a two-stage unfolding process.

The enthalpies associated with protein unfolding are often interpreted in molecular terms such as hydrogen bonds, electrostatic interactions, and hydrophobic interactions. It should be borne in mind that these interpretations are not inherent in thermodynamic quantities, which do not explicitly give information at the molecular level. Consequently, such interpretations should be scrutinized very critically.

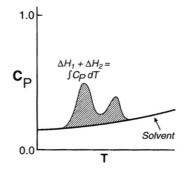

FIGURE 1-6. Schematic representation of a calorimeter scan in which the denaturation occurs in two steps. The hatched area permits the sum of the enthalpy changes to be determined, and the individual enthalpies of the unfolding reactions can be determined by a detailed analysis. As in Figure 1-5, C_P is the measured heat capacity, and T is the temperature.

1.9 REACTION ENTHALPIES

We now return to a consideration of reaction enthalpies. Because the enthalpy is a state function, it can be added and subtracted for a sequence of reactions—it does not matter how the reaction occurs or in what order. In this regard, chemical reactions can be considered as algebraic equations. For example, consider the reaction cycle below:

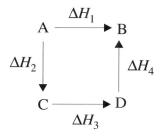

If these reactions are written sequentially, it can readily be seen how the enthalpies are related.

$$A \rightarrow C \qquad \Delta H_2$$

$$C \rightarrow D \qquad \Delta H_3$$

$$D \rightarrow B \qquad \Delta H_4$$

$$A \rightarrow B \qquad \Delta H_1 = \Delta H_2 + \Delta H_3 + \Delta H_4$$

This ability to relate enthalpies of reaction in reaction cycles in an additive fashion is often called Hess's Law, although it really is derived from thermodynamic principles as discussed. We will find that this "law" is extremely useful, as it allows determination of the enthalpy of reaction without studying a reaction directly if a sequence of reactions is known that can be added to give the desired reaction.

As an illustration, we will calculate the enthalpy of reaction for the transfer of a phosphoryl group from ATP to glucose, a very important physiological reaction catalyzed by the enzyme hexokinase.

$$\text{Glucose} + \text{ATP} \rightleftharpoons \text{ADP} + \text{Glucose-6-phosphate} \qquad (1\text{-}33)$$

The standard enthalpy changes for the hydrolysis of these four compounds are given in Table 1-1. These data are for very specific conditions: $T = 298$ K, $P = 1$ atm, pH = 7.0, pMg = 3, and an ionic strength of 0.25 M. The ionic strength is a measure of the salt concentration that takes into account the presence of both monovalent and divalent

TABLE 1-1

Reaction	ΔH_{298}° (kJ/mol)
$ATP + H_2O(\ell) \rightleftharpoons ADP + P_i$	−30.9
$ADP + H_2O(\ell) \rightleftharpoons AMP + P_i$	−28.9
$AMP + H_2O(\ell) \rightleftharpoons A + P_i$	−1.2
$G6P + H_2O(\ell) \rightleftharpoons G + P_i$	−0.5

ions ($= \frac{1}{2}\Sigma c_i z_i^2$, where c_i is the concentration of each ion, z_i is its valence, and the sum is over all of the ions present). The enthalpy change for the hexokinase reaction can easily be calculated from these data:

$$G + P_i \rightleftharpoons G6P + H_2O \qquad \Delta H_{298}^{\circ} = 0.5 \text{ kJ/mol}$$

$$ATP + H_2O \rightleftharpoons ADP + P_i \qquad \Delta H_{298}^{\circ} = -30.9 \text{ kJ/mol}$$

$$G + ATP \rightleftharpoons G6P + ADP \qquad \Delta H_{298}^{\circ} = -30.4 \text{ kJ/mol}$$

The ability to calculate thermodynamic quantities for biochemical reactions that have not yet been studied is very useful. Even if data are not available to deal with the reaction of specific interest, very often data are available for closely related reactions. Appendix 2 contains a tabulation of ΔH_{298}° for some biochemical reactions.

The enthalpy change associated with the hexokinase reaction could also be derived from the heats of formation in the table in the appendix:

$$\Delta H = H_{f,ADP}^{\circ} + H_{f,G6P}^{\circ} - H_{f,ATP}^{\circ} - H_{f,G}^{\circ}$$

$$\Delta H = -2000.2 - 2279.1 + 2981.8 + 1267.1 = -30.4 \text{ kJ/mol}$$

In point of fact, the heats of formation are usually derived from measured heats of reaction as these are the primary experimental data.

A source of potential confusion is the practice of reporting enthalpies of reaction as "per mole." There is no ambiguity for the hexokinase reaction as written above. However, in many cases, the stoichiometric coefficients for reactants and products differ. For example, the reaction catalyzed by the enzyme myokinase is

$$2 \text{ ADP} \rightleftharpoons ATP + AMP \tag{1-34}$$

Even though 2 moles of ADP are used the reaction enthalpy is referred to as "per mole." The reaction enthalpy is always given as "per mole of reaction as it is written."

It is important, therefore, that the equation for the reaction under consideration be explicitly stated. The myokinase reaction could be written as

$$ADP \rightleftharpoons \tfrac{1}{2} ATP + \tfrac{1}{2} AMP \tag{1-35}$$

In this case, the reaction enthalpy per mole would be one-half of that reported for Eq. 1-34.

1.10   TEMPERATURE DEPENDENCE OF THE REACTION ENTHALPY

In principle, the enthalpy changes as the pressure and temperature change. We will not worry about the dependence of the enthalpy on pressure, as it is usually very small for reactions in condensed phases. The temperature dependence of the enthalpy is given by Eq. 1-27. This can be used directly to determine the temperature dependence of reaction enthalpies. If we assume the standard state enthalpy is known for each reactant, then the temperature dependence of the enthalpy for each reactant, i, is

$$H_{T,i} = H_{298,i}^{\circ} + \int_{298}^{T} C_{P,i}\, dT \tag{1-36}$$

If we apply this relationship to the reaction enthalpy for the generalized reaction of Eq. 1-25, we obtain the following:

$$\Delta H_T = cH_{T,C} + dH_{T,D} - aH_{T,A} - bH_{T,B}$$

$$\Delta H_T = \Delta H_{298}^{\circ} + \int_{298}^{T} \Delta C_P\, dT$$

with

$$\Delta H_{298}^{\circ} = cH_{298,C}^{\circ} + dH_{298,D}^{\circ} - aH_{298,A}^{\circ} - bH_{298,B}^{\circ}$$

and

$$\Delta C_P = cC_{P,C} + dC_{P,D} - aC_{P,A} - bC_{P,B}$$

More generally,

$$\Delta H_T = \Delta H_{T_0} + \int_{T_0}^{T} \Delta C_P\, dT \tag{1-37}$$

Equation 1-37 is known as Kirchhoff's Law. It can also be stated in differential form:

$$d\ \Delta H/dT = \Delta C_P \tag{1-38}$$

It is important to remember that this discussion of the temperature dependence of the reaction enthalpy assumes that the pressure is constant.

The conclusion of these considerations of reaction enthalpies is that available tabulations are often sufficient to calculate the reaction enthalpy of many biological reactions. Moreover, if this is done at a standard temperature, the reaction enthalpy at other temperatures can be calculated if appropriate information about the heat capacities is known or estimated. For most chemical reactions of biological interest, the temperature dependence of the reaction enthalpy is small. On the other hand, for processes such as protein folding and unfolding, the temperature dependence is often significant and must be taken into account in data analysis and thermodynamic calculations. This will be discussed further in Chapter 3.

The first law of thermodynamics, namely, the definition of energy and its conservation, is obviously of great importance in understanding the nature of chemical reactions. As we shall see, however, the first law is not sufficient to understand what determines chemical equilibria.

REFERENCES

1. I. Tinoco, Jr., K. Sauer, and J. C. Wang, *Physical Chemistry: Principles and Applications to the Biological Sciences, 3rd* edition, Prentice Hall, Englewood Cliffs, NJ, 1995.

2. D. Eisenberg and D. Crothers, *Physical Chemistry with Applications to the Life Sciences*, Benjamin/Cummings, Menlo Park, CA, 1979.

3. *The NBS Tables of Thermodynamic Properties*, D. D. Wagman et al., eds., *J. Phys. Chem. Ref. Data, 11, Suppl. 2*, 1982.

4. D. R. Stull, E. F. Westrum, Jr., and G. C. Sinke, *The Chemical Thermodynamics of Organic Compounds*, Wiley, New York, 1969.

5. R. A. Alberty, *Arch. Biochem. Biophys.* **353**, 116 (1998).

PROBLEMS

1-1. When a gas expands rapidly through a valve, you often feel the valve get colder. This is an adiabatic expansion ($q = 0$). Calculate the decrease in temperature of 1.0 mole of ideal gas as it is expanded from 0.20 to 1.00 liter under the conditions given below. Assume a constant volume molar heat capacity, C_V, of $\frac{3}{2}R$. Note that the energy, E, of an ideal gas depends only on the temperature: It is independent of the volume of the system.

    A. The expansion is irreversible with an external pressure of 1 atmosphere and an initial temperature of 300 K.

    B. The expansion is reversible with an initial temperature of 300 K.

    C. Calculate ΔE for the changes in state described in parts A and B.

D. Assume the expansion is carried out *isothermally* at 300 K, rather than adiabatically. Calculate the work done if the expansion is carried out irreversibly with an external pressure of 1.0 atmosphere.

E. Calculate the work done if the isothermal expansion is carried out reversibly.

F. Calculate q and ΔE for the changes in state described in parts D and E.

1-2. A. Calculate the enthalpy change for the conversion of glucose $[C_6H_{12}O_6(s)]$ and oxygen $[O_2(g)]$ to $CO_2(aq)$ and $H_2O(\ell)$ under standard conditions. The standard enthalpies of formation of glucose(s), $CO_2(aq)$, and $H_2O(\ell)$ are −304.3, −98.7, and −68.3 kcal/mol, respectively.

B. When organisms metabolize glucose, approximately 50% of the energy available is utilized for chemical and mechanical work. Assume 25% of the total energy from eating one mole of glucose can be utilized to climb a mountain. How high a mountain can a 70 kg person climb?

1-3. Calculate the enthalpy change for the oxidation of pyruvic acid to acetic acid under standard conditions.

$$2\ CH_3COCOOH(\ell) + O_2(g) \rightarrow 2\ CH_3COOH(\ell) + 2\ CO_2(g)$$

The heats of combustion of pyruvic acid and acetic acid under standard conditions are −227 kcal/mol and −207 kcal/mol, respectively. Heats of combustion are determined by reacting pyruvic or acetic acid with $O_2(g)$ to give $H_2O(\ell)$ and $CO_2(g)$. *Hint*: First write balanced chemical equations for the combustion processes.

1-4. Calculate the amount of water (in liters) that would have to be vaporized at 40°C (approximately body temperature) to expend the 2.5×10^6 calories of heat generated by a person in one day (commonly called sweating). The heat of vaporization of water at this temperature is 574 cal/g. We normally do not sweat that much. What's wrong with this calculation? If 1% of the energy produced as heat could be utilized as mechanical work, how large a weight could be lifted 1 meter?

1-5. A. One hundred milliliters of 0.200 M ATP is mixed with an ATPase in a Dewar at 298 K, 1 atm, pH 7.0, pMg 3.0, and 0.25 M ionic strength. The temperature of the solution increases 1.48 K. What is $\Delta H°$ for the hydrolysis of ATP to adenosine 5′-diphosphate (ADP) and phosphate? Assume the heat capacity of the system is 418 J/K.

B. The hydrolysis reaction can be written as

$$ATP + H_2O \rightleftharpoons ADP + P_i$$

Under the same conditions, the hydrolysis of ADP,

$$ADP + H_2O \rightleftharpoons AMP + P_i$$

has a heat of reaction, $\Delta H°$, of -28.9 kJ/mol. Under the same conditions, calculate $\Delta H°$ for the adenylate kinase reaction:

$$2 \, ADP \rightleftharpoons AMP + ATP$$

1-6. The alcohol dehydrogenase reaction,

$$NAD + Ethanol \rightleftharpoons NADH + Acetaldehyde$$

removes ethanol from the blood. Use the enthalpies of formation in Appendix 1 to calculate $\Delta H°$ for this reaction. If 10.0 g of ethanol (a generous martini) is completely converted to acetaldehyde by this reaction, how much heat is produced or consumed?

CHAPTER 2

Entropy and Free Energy

2.1 INTRODUCTION

At the outset, we indicated the primary objective of our discussion of thermodynamics is to understand chemical equilibrium in thermodynamic terms. Based on our discussion thus far, one possible conclusion is that chemical equilibria are governed by energy considerations and that the system will always proceed to the lowest energy state. This idea can be discarded quite quickly, as we know some spontaneous reactions produce heat and some require heat. For example, the hydrolysis of ATP releases heat, $\Delta H^{\circ}_{298} = -30.9$ kJ/mol, whereas ATP and AMP are formed when ADP is mixed with myokinase, yet $\Delta H^{\circ}_{298} = +2.0$ kJ/mol under identical conditions. The conversion of liquids to gases requires heat, that is, ΔH is positive, even at temperatures above the boiling point. Clearly the lowest energy state is not necessarily the most stable state.

What factor is missing? (At this point, traditional treatments of thermodynamics launch into a discussion of heat engines, a topic we will avoid.) The missing ingredient is consideration of the probability of a given state. As a very simple illustration, consider three balls of equal size that are numbered 1, 2, 3. These balls can be arranged sequentially in six different ways:

$$123 \quad 132 \quad 213 \quad 231 \quad 312 \quad 321$$

The energy state of all of these arrangements is the same, yet it is obvious that the probability of the balls being in sequence (123) is 1/6, whereas the probability of the balls being out of sequence is 5/6. In other words, the probability of a disordered state is much greater than the probability of an ordered state because a larger number of arrangements of the balls exists in the disordered state.

Molecular examples of this phenomenon can readily be found. A gas expands spontaneously into a vacuum even though the energy state of the gas does not change. This occurs because the larger volume has more positions available for molecules, so a greater number of arrangements, or more technically *microstates*, of molecules are possible. Clearly, probability considerations are not sufficient by themselves. If this were the case, the stable state of matter would always be a gas. We know solids and liquids are stable under appropriate conditions because they are energetically favored; that is, interactions between atoms and molecules result in a lower energy state. The real situation must involve a balance between energy and probability. This is a qualitative statement of what determines the equilibrium state of a system, but we will be able to be much more quantitative than this.

The second law states that disordered states are more probable than ordered states. This is done by defining a new state function, *entropy*, which is a measure of the disorder (or probability) of a state. Thermodynamics does not require this interpretation of the entropy, which is quasi-molecular. However, this is a much more intuitive way of understanding entropy than utilizing the traditional concept of heat engines. The more disordered a state, or the larger the number of available microstates, the higher the entropy. We already can see a glimmer of how the equilibrium state might be determined. At constant entropy, the energy should be minimized whereas at constant energy, the entropy should be maximized. We will return to this topic a little later. First, we will define the entropy quantitatively.

2.2   STATEMENT OF THE SECOND LAW

A more formal statement of the second law is to define a new state function, the entropy, S, by the equation

$$dS = \frac{dq_{rev}}{T} \tag{2-1}$$

or

$$\Delta S = \int \frac{dq_{rev}}{T} \tag{2-2}$$

The temperature scale in this definition is Kelvin. This definition is not as straightforward as that for the energy. Note that this definition requires a reversible heat change, q_{rev} or dq_{rev}, yet entropy is a state function. At first glance, this seems quite paradoxical. The meaning of this is that the entropy change must be calculated by finding a reversible path. However, all reversible paths give the same entropy change, and the calculated entropy change is correct even if the actual change in state is carried out irreversibly. Although this appears to be somewhat confusing, consideration of some examples will help in understanding this concept.

The second law also includes important considerations about entropy: For a reversible change in state, the entropy of the universe is constant, whereas for an irreversible change in state, the entropy of the universe increases.

Again, the second law cannot be proved, but we can demonstrate that without this law, events could transpire that are contrary to our everyday experience. Two examples are given below.

Without the second law, a gas could spontaneously compress! Let's illustrate this by considering the isothermal expansion of an ideal gas, V_1 to V_2 with constant T. The entropy change is

$$\Delta S = \int (dq_{rev}/T) = (1/T) \int dq_{rev} = q_{rev}/T \tag{2-3}$$

For the isothermal expansion of an ideal gas, $\Delta E = 0$. (Because of the definition of an ideal gas, the energy, E, is determined by the temperature only and does not depend on the volume and pressure.) Therefore,

$$q_{rev} = -w_{rev} = \int_{V_1}^{V_2} P\, dV = \int_{V_1}^{V_2} (nRT/V)\, dV = nRT \ln(V_2/V_1) \tag{2-4}$$

and

$$\Delta S = nR\, \ln(V_2/V_1) \tag{2-5}$$

For an expansion, $V_2 > V_1$, and $\Delta S > 0$. For a compression, $V_2 < V_1$, and $\Delta S < 0$. The second law does not prohibit this situation, as we are considering the entropy change for the system, not the universe. This result is in accord with the intuitive interpretation of entropy previously discussed: A larger volume has more positions for the gas to occupy and consequently a higher entropy.

We must now consider what happens to the entropy of the surroundings. For a reversible change, $\Delta S = -q_{rev}/T$, and the entropy of the universe is the sum of the entropy change for the system and that for the surroundings: $\Delta S = q_{rev}/T - q_{rev}/T = 0$, which is consistent with the second law. However, for an irreversible change the situation is different. Let's make this irreversible by setting the external pressure equal to zero during the change in state. In that case, $w = 0$ and since $\Delta E = 0$, $q = 0$. Therefore, no heat is lost by the surroundings. The entropy change for the universe is

$$\Delta S = \Delta S_{gas} + \Delta S_{surr} = nR\, \ln(V_2/V_1) + 0$$

The second law says that the entropy change of the universe must be greater than zero, which requires that $V_2 > V_1$. In other words, a spontaneous compression cannot occur. This is not required by the first law.

As a second example, consider two blocks at different temperatures, T_h and T_c, where h and c designate hot and cold so $T_h > T_c$. We will put the blocks together so that heat is transferred from the hot block to the cold block. The entropy changes in the two blocks are given by

$$dS_c = dq/T_c \quad \text{and} \quad dS_h = -dq/T_h$$

If the two blocks are considered to be the universe, the entropy change of the universe is

$$dS_c + dS_h = dq\,(1/T_c - 1/T_h) > 0$$

As predicted by the second law, the entropy of the universe increases. What if the heat flows from the cold block to the hot block? Then the sign of the heat change is reversed and

$$dS_c + dS_h = dq \, (1/T_h - 1/T_c) < 0$$

This predicts the entropy of the universe would decrease, which is impossible according to the second law. Thus, heat cannot flow from the cold bar to the hot bar. This is not prohibited by the first law.

Exceptions to the second law can be used to create perpetual motion machines, which we know are impossible. These are sometimes called perpetual motion machines of the second kind, whereas perpetual motion machines created by exceptions to the first law are called perpetual motion machines of the first kind.

2.3 CALCULATION OF THE ENTROPY

A reversible path must always be found to calculate the entropy change. At constant volume, the relationship

$$dq_{rev} = nC_V \, dT \tag{2-6}$$

can be used, whereas at constant pressure

$$dq_{rev} = nC_P \, dT \tag{2-7}$$

or

$$\Delta S = \int nC_P \, dT/T \tag{2-8}$$

Entropy changes for phase changes are particularly easy to calculate since they occur at constant temperature and pressure. At constant temperature and pressure,

$$\Delta S = q_{rev}/T = \Delta H/T \tag{2-9}$$

For example, for the process

$$\text{Benzene(s, 1 atm, 279 K)} \rightarrow \text{Benzene}(\ell, 1 \text{ atm, 279 K}) \tag{2-10}$$

$\Delta H = 2380$ cal/mol, and $\Delta S = 2380/279 = 8.53$ cal/(mol·K) = 8.53 eu. Here 1 cal/(mol·K) is defined as an entropy unit, eu. The entropy change for the reverse process is -8.53 eu. Note that the reverse process is not prohibited by the second law since it is the entropy change for the system, not the universe. Also note that, as expected, going from a solid to a liquid involves a positive entropy change since the liquid state is more disordered than the solid.

While it is easy to state that the entropy change can be calculated for an irreversible process by finding a reversible way of going from the initial state to the final state, this is not always easy to do. This is usually not a matter of great consequence for our con-

siderations, but we will consider one example to illustrate the process. Let us determine the entropy change for the following process:

$$H_2O(\ell, 298 \text{ K}, 1 \text{ atm}) \rightarrow H_2O(g, 298 \text{ K}, 1 \text{ atm}) \qquad (2\text{-}11)$$

This change is not a reversible change in state, as we know that the normal boiling
point of water at 1 atmosphere is 373 K. A possible reversible cycle that would go
from the initial state to the final state at constant pressure is

$$H_2O(\ell, 298 \text{ K}, 1 \text{ atm}) \rightarrow H_2O(g, 298 \text{ K}, 1 \text{ atm}) \qquad (2\text{-}12)$$
$$\downarrow \qquad\qquad\qquad \uparrow$$
$$H_2O(\ell, 373 \text{ K}, 1 \text{ atm}) \rightarrow H_2O(g, 373 \text{ K}, 1 \text{ atm}).$$

The entropy change for the bottom process, which is reversible, is simply $\Delta H/T = 9710/373 = 26$ cal/mol·K). The entropy change for the left-hand side of the square is
(Eq. 2-8)

$$\Delta S = C_P \ln(T_2/T_1) = 18 \ln(373/298) = 4 \text{ cal/(mol·K)}$$

and for the right-hand side of the square is

$$\Delta S = C_P \ln(T_2/T_1) = 8.0 \ln(298/373) = -1.8 \text{ cal/(mol·K)}$$

The entropy changes for these three reversible processes can be added to give the entropy change for the change in state given in Eq. 2-11: 28 cal/(mol·K).

An alternative reversible path that can be constructed lowers the pressure to the
equilibrium vapor pressure of water at 298 K. The corresponding constant temperature
cycle is

$$H_2O(\ell, 298 \text{ K}, 1 \text{ atm}) \rightarrow H_2O(g, 298 \text{ K}, 1 \text{ atm}) \qquad (2\text{-}13)$$
$$\downarrow \qquad\qquad\qquad \uparrow$$
$$H_2O(\ell, 298 \text{ K}, 0.0313 \text{ atm}) \rightarrow H_2O(g, 298 \text{ K}, 0.0313 \text{ atm})$$

In this case, we would have to calculate the change in entropy as the pressure is lowered and raised. This can easily be done but is beyond the scope of this presentation
of thermodynamics. The point of this exercise is to illustrate how entropy changes can
be calculated for irreversible as well as reversible processes and multiple reversible
processes can be found.

In principle, the entropy can be calculated from statistical considerations.
Boltzmann derived a relationship between the entropy and the number of microstates,
N:

$$S = k_B \ln N \qquad (2\text{-}14)$$

where k_B is Boltzmann's constant, 1.38×10^{-23} J/K. It is rarely possible to determine the number of microstates although the number of microstates could be calculated from Eq. 2-14 if the entropy is known. For a simple case, such as the three numbered balls with which we started our discussion of the second law, this calculation can readily be done. The disordered system has 3! microstates and an entropy of 1.51×10^{-23} J/K. Any ordered sequence—for example 1, 2, 3, —has only 1 microstate, so $N = 1$, and $S = 0$. Since the disordered state has a higher entropy, this predicts that the balls will spontaneously disorder, and that an ordered state is extremely unlikely.

2.4 THIRD LAW OF THERMODYNAMICS

We will not dwell on the third law as the details are of little consequence in biology. The important fact for us is that the third law establishes a zero for the entropy scale. Unlike the energy, entropy has an absolute scale. The third law can be stated as follows: The entropy of perfect crystals of all pure elements and compounds is zero at absolute zero. The tricky points of this law are the meanings of "perfect" and "pure," but we will not discuss this in detail. It is worth noting that a perfect crystal has one microstate, and therefore an entropy of zero according to Eq. 2-14.

The absolute standard entropy can be determined from measurements of the temperature dependence of the heat capacity using the relationship

$$S^{\circ}_{298} = \int_{0}^{298} C_P \, dT/T \tag{2-15}$$

Here the entropy at absolute zero has been assumed to be zero in accord with the third law. A plot of C_P/T versus T gives a curve such as that in Figure 2-1. The area under the curve is the absolute standard entropy. Tables of S°_{298} are readily available (cf.

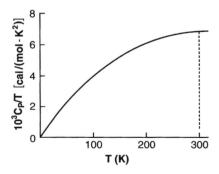

FIGURE 2-1. A plot of the constant-pressure heat capacity divided by the temperature, C_P/T, versus the temperature, T, for graphite. The absolute entropy of graphite at 300 K is the area under the curve up to the dashed line (Eq. 2-15). If phase changes occur as the temperature is changed, the entropy changes associated with the phase changes must be added to the area under the curve of the C_P/T versus T plot.

Refs. 3 and 4, Chapter 1). The entropy at temperatures other than 298 K can be calculated from the relationship

$$S_T^\circ = S_{298}^\circ + \int_{298}^{T} C_P \, dT/T \tag{2-16}$$

For a chemical reaction,

$$\Delta S_T = \Delta S_{298}^\circ + \int_{298}^{T} \Delta C_P \, dT/T \tag{2-17}$$

These relationships are analogous to those used for the enthalpy. Indeed, entropies of reactions can be calculated in a similar fashion to enthalpy changes.

2.5 MOLECULAR INTERPRETATION OF ENTROPY

We will now consider a few examples of absolute entropies and entropy changes for chemical reactions, and how they might be interpreted in molecular terms. The absolute entropies for water as a solid, liquid, and gas at 273 K and 1 atmosphere, are 41.0, 63.2, and 188.3 J/(mol·K), respectively. The molecular interpretation of these numbers is straightforward, namely, that solid is more ordered than liquid, which is more ordered than gas; therefore, the solid has the lowest entropy and the gas the highest.

Standard entropy changes for some chemical reactions are given in Table 2-1. As expected, the entropy change is negative for the first two reactions in the gas phase as the number of moles of reactants is greater than the number of moles of products. Of course, these reactions could have been written in the opposite direction. The entropy change would then have the opposite sign. At first glance, the result for the third reaction in the table is surprising, as the number of moles of reactants is greater than the number of moles of products. This results because the solvent must be included in the molecular interpretation of the observed entropy change. Ions in water interact strongly with water so that highly ordered water molecules exist around the ions.

TABLE 2-1

Reaction	ΔS_{298}° [J/(mol·K)]
$H(g) + H(g) \rightleftharpoons H_2(g)$	−98.7
$2\,H_2(g) + O_2(g) \rightleftharpoons 2\,H_2O(g)$	−88.9
$H^+(aq) + OH^-(aq) \rightleftharpoons H_2O(\ell)$	+80.7
Cytidine 2′-monophosphate + Ribonuclease A $\rightleftharpoons$ Enzyme–inhibitor complex	−54[a]

[a]H. Naghibi, A. Tamura, and J. M. Sturtevant, *Proc. Natl. Acad. Sci. USA* **92**, 5597 (1995).

When the neutral species is formed, these highly ordered water molecules become less ordered. Thus, the entropy change for water is very positive, much more positive than the expected entropy decrease for H^+ and OH^- when they form water. This simple example indicates that considerable care must be exercised in interpreting entropy changes for chemical reactions in condensed media. The entropy of the entire system under consideration must be taken into account. The final entry is for the binding of a ligand to an enzyme. In this case, no reasonable interpretation of the entropy change is possible as three factors come into play: the loss in entropy as two reactants become a single entity, the changes in the structure of water, and structural changes in the protein. The fact that the entropy change is comparable to the value expected for the combination of two molecules to produce one molecule is simply fortuitous. Extreme caution should be exercised in making molecular interpretations of thermodynamic changes in complex systems.

We have established all of the thermodynamic principles necessary to discuss chemical equilibrium. We will now apply these principles to develop a general framework for dealing with chemical reactions.

2.6 FREE ENERGY

In a sense we have reached our goal. We have developed thermodynamic criteria for the occurrence of reversible and irreversible (spontaneous) processes, namely, for the universe, $\Delta S = 0$ for reversible processes and must be greater than zero for irreversible processes. Unfortunately, this is not terribly useful, as we are interested in what is happening in the system and require criteria that are easily applicable to chemical reactions. This can be achieved by defining a new thermodynamic state function, the Gibbs free energy:

$$G = H - TS \qquad (2\text{-}18)$$

(J. Willard Gibbs developed the science of thermodynamics virtually single handed around the turn of the century at Yale University. To this day, his collected works are prized possessions in the libraries of people seriously interested in thermodynamics.)

At constant temperature,

$$\Delta G = \Delta H - T\,\Delta S$$

$$= q_P - T\,\Delta S = q_P - q_{\text{rev}} \qquad (2\text{-}19)$$

For a reversible process, $q_P = q_{\text{rev}}$ so $\Delta G = 0$. For irreversible processes, the situation is a bit more complex as the sign of the heat change must be considered. For an endothermic process, $q_P < q_{\text{rev}}$ so $\Delta G < 0$. For an exothermic process, the absolute value of q_P is greater than the absolute value of q_{rev} so again, $\Delta G < 0$. This is the free energy change for the system and does not involve the surroundings. We now have developed criteria that tell us if a process occurs spontaneously. If $\Delta G < 0$, the change in state occurs spontaneously, whereas if $\Delta G > 0$, the reverse change in state occurs spontaneously. If $\Delta G = 0$, the system is at equilibrium (at constant pressure and temperature).

As with the energy and enthalpy, only differences in free energy can be measured. Consequently, the zero of the free energy scale is arbitrary. As for the enthalpy, the zero of the scale is taken as the elements in their stable state at 1 atmosphere and 298 K. Again, analogous to the enthalpy, tables of the free energies of formation of compounds are available so that free energy changes for chemical reactions can be calculated (cf. Refs. 3–5 in Chapter 1). Referring back to the reaction in Eq. 1-25,

$$\Delta G_{298}^{\circ} = cG_{298,C}^{\circ} + dG_{298,D}^{\circ} - aG_{298,A}^{\circ} - bG_{298,B}^{\circ} \qquad (2\text{-}20)$$

where the free energies on the right-hand side of the equation are free energies per mole. We will discuss the temperature dependence of the free energy a bit later, but it is useful to remember that at constant temperature and pressure,

$$\Delta G = \Delta H - T\,\Delta S \qquad (2\text{-}21)$$

Standard state free energy changes are available for biochemical reactions although comprehensive tabulations do not exist, and the definition of "standard state" is more complex than in the gas phase, as discussed previously. Standard state free energies of formation for some substances of biochemical interest are given in Appendix 1.

As an illustration of the concept of free energy, consider the conversion of liquid water to steam:

$$H_2O(\ell) \rightleftharpoons H_2O(g) \qquad (2\text{-}22)$$

At the boiling point, $\Delta H = 9710$ cal/mol and $\Delta S = 26$ eu. Therefore,

$$\Delta G = 9710 - 26T \qquad (2\text{-}23)$$

At equilibrium, $\Delta G = 0$ and Eq. 2-23 gives $T = 373$ K, the normal boiling point of water. If $T > 373$ K, $\Delta G < 0$, and the change in state is spontaneous, whereas if $T < 373$ K, $\Delta G > 0$, and the reverse process, condensation, occurs spontaneously.

Calculation of the free energy for a change in state is not always straightforward. Because the entropy is part of the definition of the free energy, a reversible process must always be found to calculate the free energy change. The free energy change, of course, is the same regardless of how the change in state is accomplished since the free energy is a state function. As a simple example, again consider the change in state in Eq. 2-11. This is not a reversible process since the two states are not in equilibrium. This also is not a spontaneous change in state so that $\Delta G > 0$. In order to calculate the value of ΔG, we must think of a reversible path for carrying out the change in state. The two cycles in Eqs. 2-12 and 2-13 again can be used. In both cases, the bottom reaction is an equilibrium process, so $\Delta G = 0$. To calculate ΔG for the top reaction, we only need to add up the ΔG values for the vertical processes. These can be calculated from the temperature (upper cycle) and pressure (lower cycle) dependence of G. Such functional dependencies will be considered shortly. It will be a useful exercise for the reader to carry out the complete calculations.

2.7 CHEMICAL EQUILIBRIA

Although we now have developed criteria for deciding whether or not a process will occur spontaneously, they are not sufficient for consideration of chemical reactions. We know that chemical reactions are generally not "all or nothing" processes; instead, an equilibrium state is reached where both reactants and products are present. We will now derive a quantitative relationship between the free energy change and the concentrations of reactants and products. We will do this in detail for the simple case of ideal gases, and by analogy for reactions in liquids.

The starting point for the derivation is the definition of free energy and its total derivative:

$$G = H - TS = E + PV - TS \tag{2-24}$$

$$dG = dE + P\,dV + V\,dP - T\,dS - S\,dT \tag{2-25}$$

Since $dE = dq + dw = T\,dS - P\,dV$ for a reversible process,

$$dG = V\,dP - S\,dT \tag{2-26}$$

(Although this relationship was derived for a reversible process, it is also valid for an irreversible process.) Let us now consider a chemical reaction of ideal gases at constant temperature. For one mole of each gas component,

$$dG = V\,dP = RT\,dP/P \tag{2-27}$$

We will refer all of our calculations to a pressure of 1 atmosphere for each component. In thermodynamic terms, we have selected 1 atmosphere as our *standard state*. If Eq. 2-27 is now integrated from $P_0 = 1$ atmosphere to P,

$$dG = RT\,dP/P$$

or

$$G = G^\circ + RT\ln(P/P_0) = G^\circ + RT\ln P \tag{2-28}$$

We will now return to our prototype reaction, Eq. 1-25, and calculate the free energy change. The partial pressures of the reactants are given in parentheses:

$$aA(P_A) + bB(P_B) \rightleftharpoons cC(P_C) + dD(P_D) \tag{2-29}$$

$$\Delta G = cG_C + dG_D - aG_A - bG_B = cG_C^\circ + dG_D^\circ - aG_A^\circ - bG_B^\circ + cRT \ln P_C + dRT \ln P_D$$

$$- aRT \ln P_A - bRT \ln P_B$$

or

$$\Delta G = \Delta G^\circ + RT \ln\left(\frac{P_C^c P_D^d}{P_A^a P_B^b}\right) \tag{2-30}$$

The ΔG is the free energy for the reaction in Eq. 2-29 when the system is not at equilibrium. At equilibrium, at constant temperature and pressure, $\Delta G = 0$, and Eq. 2-30 becomes

$$\Delta G^\circ = - RT \ln\left(\frac{P_C^c P_D^d}{P_A^a P_B^b}\right)_e = - RT \ln K \tag{2-31}$$

Here the subscript e has been used to designate equilibrium and K is the equilibrium constant.

We now have a quantitative relationship between the partial pressures of the reactants and the standard free energy change, ΔG°. The standard free energy change is a constant at a given temperature and pressure but will vary as the temperature and pressure change. If $\Delta G^\circ < 0$, then K > 1, whereas if $\Delta G^\circ > 0$, K < 1. A common mistake is to confuse the free energy change with the standard free energy change. The free energy change is always equal to zero at equilibrium and can be calculated from Eq. 2-30 when not at equilibrium. The standard free energy change is a constant representing the hypothetical reaction with all of the reactants and products at a pressure of 1 atmosphere. It is equal to zero only if the equilibrium constant fortuitously is 1.

Biological reactions do not occur in the gas phase. What about free energy in solutions? Conceptually there is no difference. The molar free energy at constant temperature and pressure can be written as

$$G = G^\circ + RT \ln(c/c_0) \tag{2-32}$$

where c is the concentration and c_0 is the standard state concentration. A more correct treatment would define the molar free energy as

$$G = G^\circ + RT \ln a \tag{2-33}$$

where a is the thermodynamic activity and is dimensionless. However, the thermodynamic activity can be written as a product of an activity coefficient and the concentration. The activity coefficient can be included in the standard free energy, $G°$, which gives rise to Eq. 2-32. We need not worry about this as long as the solution conditions are clearly defined with respect to salt concentration, pH, etc. The reason it is not of great concern is that all of the aforementioned complications can be included in the standard free energy change since, in practice, the standard free energy is determined by measuring the equilibrium constant under defined conditions.

Finally, we should note that the free energy per mole at constant temperature and pressure is called the chemical potential, μ, in more sophisticated treatments of thermodynamics, but there is no need to introduce this terminology here.

If we take the standard state as 1 mole/ liter, then the results parallel to Eqs. 2-28, 2-30, and 2-31 are

$$G = G° + RT \ln c \tag{2-34}$$

$$\Delta G = \Delta G° + RT \ln\left(\frac{c_C^c c_D^d}{c_A^a c_B^b}\right) \tag{2-35}$$

$$\Delta G° = - RT \ln\left(\frac{c_C^c c_D^d}{c_A^a c_B^b}\right)_e = - RT \ln K \tag{2-36}$$

Equations 2-34 to 2-36 summarize the thermodynamic relationships necessary to discuss chemical equilibria.

Note that, strictly speaking, the equilibrium constant is dimensionless as all of the concentrations are ratios, the actual concentration divided by the standard state concentration. However, practically speaking, it is preferable to report equilibrium constants with the dimensions implied by the ratio of concentrations in Eq. 2-36. The equilibrium constant is determined experimentally by measuring concentrations, and attributing dimensions to this constant assures that the correct ratio of concentrations is considered and that the standard state is precisely defined.

Consideration of the free energy also allows us to assess how the energy and entropy are balanced to achieve the final equilibrium state. Since $\Delta G = \Delta H - T\Delta S$ at constant T and P, it can be seen that a change in state is spontaneous if the enthalpy change is very negative and/or the entropy change is very positive. Even if the enthalpy change is unfavorable (positive), the change in state will be spontaneous if the $T\Delta S$ term is very positive. Similarly, even if the entropy change is unfavorable ($T\Delta S$ very negative), the change in state will be spontaneous if the enthalpy change is sufficiently negative. Thus, the final equilibrium achieved is a balance between the enthalpy (ΔH) and the entropy ($T\Delta S$).

2.8 PRESSURE AND TEMPERATURE DEPENDENCE OF THE FREE ENERGY

We will now return to the pressure and temperature dependence of the free energy. At constant temperature, the pressure dependence of the free energy follows directly from Eq. 2-26, namely,

$$dG = V \, dP \tag{2-37}$$

This equation can be integrated if the pressure dependence of the volume is known, as for an ideal gas. For a chemical reaction, Eq. 2-37 can be rewritten as

$$d \, \Delta G = \Delta V \, dP \tag{2-38}$$

where ΔV is the difference in volume between the products and reactants. The pressure dependence of the equilibrium constant at constant temperature follows directly,

$$d \Delta G^\circ = -RT \, d \ln K = \Delta V \, dP \tag{2-39}$$

or

$$\frac{d \ln K}{dP} = -\frac{\Delta V}{RT} \tag{2-40}$$

For most chemical reactions, the pressure dependence of the equilibrium constant is quite small so that it is not often considered in biological systems.

Equilibrium constants, however, frequently vary significantly with temperature. At constant pressure, Eq. 2-26 gives

$$dG = -S \, dT \tag{2-41}$$

or

$$d \, \Delta G = -\Delta S \, dT \tag{2-42}$$

Returning to the basic definition of free energy at constant temperature and pressure,

$$\Delta G = \Delta H - T \, \Delta S = \Delta H + T(d \, \Delta G/dT)$$

This equation can be divided by T^2 and rearranged as follows:

$$-\Delta G/T^2 + (d \, \Delta G/dT)/T = -\Delta H/T^2$$

$$\frac{d(\Delta G/T)}{dT} = -\frac{\Delta H}{T^2} \tag{2-43}$$

Equation 2-43 is an important thermodynamic relationship describing the temperature dependence of the free energy at constant pressure and is called the Gibbs–Helmholtz equation. The temperature dependence of the equilibrium constant follows directly:

$$\frac{d(\Delta G^\circ/T)}{dT} = -R\frac{d \ln K}{dT}$$

or

$$\frac{d \ln K}{dT} = \frac{\Delta H^\circ}{RT^2} \tag{2-44}$$

If ΔH° is independent of temperature, Eq. 2-44 can easily be integrated:

$$d \ln K = \int_{T_1}^{T_2} \frac{\Delta H^\circ dT}{RT^2}$$

$$\ln\left(\frac{K_2}{K_1}\right) = \frac{\Delta H^\circ}{R}\left(\frac{1}{T_1} - \frac{1}{T_2}\right) = \frac{(\Delta H^\circ/R)(T_2 - T_1)}{T_1 T_2} \tag{2-45}$$

When carrying out calculations, the difference between reciprocal temperatures should never be used directly as it introduces a large error. Instead the rearrangement in Eq. 2-45 should be used in which the difference between two temperatures occurs. With this equation and a knowledge of ΔH°, the equilibrium constant can be calculated at any temperature if it is known at one temperature.

What about the assumption that the standard enthalpy change is independent of temperature? This assumption is reasonable for many biological reactions if the temperature range is not too large. In some cases, the temperature dependence cannot be neglected and must be included explicitly in carrying out the integration of Eq. 2-44. The temperature dependence of the reaction enthalpy depends on the difference in heat capacities between the products and reactants as given by Eq. 1-38.

Examples of the temperature dependence of equilibrium constants are displayed in Figure 2-2. As predicted by Eq. 2-45, a plot of $\ln K$ versus $1/T$ is a straight line with a slope of $-\Delta H^\circ/R$. The data presented are for the binding of DNA to DNA binding proteins (i.e., Zn fingers). The dissociation constants for binding are quite similar for both proteins at 22°C, 1.08×10^{-9} M (WT1 protein), and 3.58×10^{-9} M (EGR1 protein). However, as indicated by the data in the figure, the standard enthalpy changes are *opposite* in sign, +6.6 kcal/mol and −6.9 kcal/mol, respectively. Consequently, the standard entropy changes are also quite different, 63.3 eu and 15.3 eu, respectively. These data indicate that there are significant differences in the binding processes, despite the similar equilibrium constants.

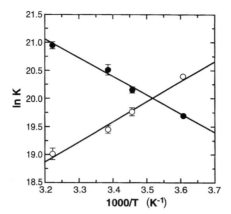

FIGURE 2-2. Temperature dependence of the equilibrium binding constant, K, for the binding of DNA to binding proteins ("zinc fingers"), WT1 (●) and EGR1 (○), to DNA. Adapted with permission from T. Hamilton, F. Borel, and P. J. Romaniuk, *Biochemistry* **37**, 2051 (1998). Copyright © 1998 American Chemical Society.

2.9 PHASE CHANGES

The criterion that $\Delta G = 0$ at equilibrium at constant temperature and pressure is quite general. For chemical reactions, this means that the free energy of the products is equal to the free energy of the reactants. For phase changes of pure substances, this means that at equilibrium the free energies of the phases are equal. If we assume two phases, A and B, Eq. 2-26 gives

$$dG_A = V_A\,dP - S_A\,dT = dG_B = V_B\,dP - S_B\,dT$$

Rearrangement gives

$$\frac{dP}{dT} = \frac{\Delta S_{BA}}{\Delta V_{BA}} \qquad (2\text{-}46)$$

where $\Delta S_{BA} = S_B - S_A$ and $\Delta V_{BA} = V_B - V_A$. (All of these quantities are assumed to be per mole for simplicity.) This equation is often referred to as the Clapeyron equation. Note that the entropy change can be written as

$$\Delta S_{BA} = \Delta H_{BA}/T \qquad (2\text{-}47)$$

Equation 2-46 gives the slope of *phase diagrams*, plots of P versus T that are useful summaries of the phase behavior of a pure substance. As an example, the phase diagram of water is given in Figure 2-3.

The lines in Figure 2-3 indicate when two phases are in equilibrium and coexist. Only one phase exists in the open areas, and when the two lines meet, three phases are in equilibrium. This is called the triple point and can only occur at a pressure of 0.006

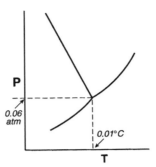

FIGURE 2-3. Schematic representation of the phase diagram of water with pressure, P, and temperature, T, as variables. The phase diagram is not to scale and is incomplete, as several different phases of solid water are known. If volume is included as a variable, a three-dimensional phase diagram can be constructed.

atmosphere and 0.01°C. The slopes of the lines are given by Eq. 2-46. The number of phases that can coexist is governed by the phase rule,

$$f = c - p + 2 \tag{2-48}$$

where f is the number of degrees of freedom, c is the number of components, and p is the number of phases. This rule can be derived from the basic laws of thermodynamics. For a pure substance, the number of components is 1, so $f = 3 - p$. In the open spaces, one phase exists, so that the number of degrees of freedom is 2, which means that both T and P can be varied. Two phases coexist on the lines, which gives $f = 1$. Because only one degree of freedom exists, at a given value of P, T is fixed and vice versa. At the triple point, three phases coexist, and there are no degrees of freedom, which means that both P and T are fixed.

The study of phase diagrams is an important aspect of thermodynamics, and some interesting applications exist in biology. For example, membranes contain phospholipids. If phospholipids are mixed with water, they spontaneously form bilayer vesicles with the head groups pointing outward toward the solvent so that the interior of the bilayer is the hydrocarbon chains of the phospholipids. A schematic representation of the bilayer is given in Figure 2-4. The hydrocarbon chain can be either disordered, at high temperatures, or ordered, at low temperatures, as indicated schematically in the figure. The transformation from one form to the other behaves as a phase transition so that phase diagrams can be constructed for phospholipids. Moreover, the phase diagrams depend not only on the temperature but on the phospholipid composition of the bilayer. This phase transition has biological implications in that the fluidity of the hydrocarbon portion of the membrane strongly affects the transport and mechanical properties of the membranes. Phase diagrams for phospholipids and phospholipid–water mixtures can be constructed by a variety of methods, including scanning calorimetry and nuclear magnetic resonance.

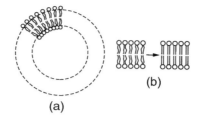

(b)

(a)

FIGURE 2-4. Schematic representation of the vesicles of phospholipids that are formed when phospholipids are suspended in water (a). Here the small circles represent the head groups and the wiggly lines the two hydrocarbon chains. A phase change can occur in the side chains from a disordered to an ordered state as sketched in (b).

2.10 ADDITIONS TO THE FREE ENERGY

As a final consideration in developing the concept of free energy, we will return to our original development of equilibrium, namely, Eq. 2-33. In arriving at this expression for the molar free energy at constant temperature and pressure, the assumption was made that only $P–V$ work occurs. This is not true in general and for many biological systems in particular. If we went all the way back to the beginning of our development of the free energy changes associated with chemical equilibria (Eq. 2-26) and derived the molar free energy at constant temperature and pressure, we would find that

$$G = G^\circ + RT \ln c + w_{max} \tag{2-49}$$

where w_{max} is the maximum (reversible) non $P–V$ work. For example, for an ion of charge z in a potential field Ψ, w_{max} is $zF\Psi$, where F is the Faraday. (We have used this relationship in Chapter 1 to calculate the work of moving an ion across a membrane with an imposed voltage. In terms of transport of an ion across a membrane, this assumes the ion is being transported from the inside to the outside with an external membrane potential of Ψ.) This *extended chemical potential* is useful for discussing ion transport across membranes, as we shall see later. It could also be used, for example, for considering molecules in a gravitational or centrifugal field. (See Ref. 1 in Chapter 1 for a detailed discussion of this concept.)

We are now ready to consider some applications of thermodynamics to biological systems in more detail.

PROBLEMS

2-1. One mole of an ideal gas initially at 300 K is expanded from 0.2 to 1.0 liter. Calculate ΔS for this change in state if it is carried out under the following conditions. Assume a constant volume heat capacity for the gas, C_V, of $\frac{3}{2}R$.

   A. The expansion is *reversible* and *isothermal.*

B. The expansion is *irreversible* and *isothermal*.

C. The expansion is *reversible* and *adiabatic* ($q = 0$).

D. The expansion is *irreversible* with an external pressure of 1 atmosphere and *adiabatic*.

2-2. The alcohol dehydrogenase reaction, which removes ethanol from your blood, proceeds according to the following reaction:

$$NAD^+ + Ethanol \rightleftharpoons NADH + Acetaldehyde$$

Under standard conditions (298 K, 1 atm, pH 7.0, pMg 3, and an ionic strength of 0.25 M), the standard enthalpies and free energies of formation of the reactants are as follows:

	$H°$ (kJ/mol)	$G°$ (kJ/mol)
NAD$^+$	−10.3	1059.1
NADH	−41.4	1120.1
Ethanol	−290.8	63.0
Acetaldehyde	−213.6	24.1

A. Calculate $\Delta G°$, $\Delta H°$, and $\Delta S°$ for the alcohol dehydrogenase reaction under standard conditions.

B. Under standard conditions, what is the equilibrium constant for this reaction? Will the equilibrium constant increase or decrease as the temperature is increased?

2-3. The equilibrium constant under standard conditions (1 atm, 298 K, pH 7.0) for the reaction catalyzed by fumarase,

$$Fumarate + H_2O \rightleftharpoons L\text{-malate}$$

is 4.00. At 310 K, the equilibrium constant is 8.00.

A. What is the standard free energy change, $\Delta G°$, for this reaction?

B. What is the standard enthalpy change, $\Delta H°$, for this reaction? Assume the standard enthalpy change is independent of temperature.

C. What is the standard entropy change, $\Delta S°$, at 298 K for this reaction?

D. If the concentration of both reactants is equal to 0.01 M, what is the free energy change at 298 K? As the reaction proceeds to equilibrium, will more fumarate or L-malate form?

E. What is the free energy change for this reaction when equilibrium is reached?

2-4. The following reaction is catalyzed by the enzyme creatine kinase:

Creatine + ATP $\rightleftharpoons$ Creatine phosphate + ADP

A. Under standard conditions (1 atm, 298 K, pH 7.0, pMg 3.0 and an ionic strength of 0.25 M) with the concentrations of all reactants equal to 10 mM, the free energy change, ΔG, for this reaction is 13.3 kJ/mol. What is the standard free energy change, $\Delta G°$, for this reaction?

B. What is the equilibrium constant for this reaction?

C. The standard enthalpies of formation for the reactants are as follows:

Creatine	−540 kJ/mol
Creatine phosphate	−1510 kJ/mol
ATP	−2982 kJ/mol
ADP	−2000 kJ/mol

What is the standard enthalpy change, $\Delta H°$, for this reaction?

D. What is the standard entropy change, $\Delta S°$, for this reaction?

2-5. A. It has been proposed that the reason ice skating works so well is that the pressure from the blades of the skates melts the ice. Consider this proposal from the viewpoint of phase equilibria. The phase change in question is

$$H_2O(s) \rightleftharpoons H_2O(\ell)$$

Assume that ΔH for this process is independent of temperature and pressure and is equal to 80 cal/g. The change in volume, ΔV, is about -9.1×10^{-5} L/g. The pressure exerted is the force per unit area. For a 180 pound person and an area for the skate blades of about 6 square inches, the pressure is 30 lb/sq. in. or about 2 atmospheres. With this information, calculate the decrease in the melting temperature of ice caused by the skate blades. (Note that 1 cal = 0.04129 L·atm.) Is this a good explanation for why ice skating works?

B. A more efficient way of melting ice is to add an inert compound such as urea. (We will avoid salt to save our cars.) The extent to which the freezing point is lowered can be calculated by noting that the molar free energy of water must be the same in the solid ice and the urea solution. The molar free energy of water in the urea solution can be approximated as $G°_{liquid} + RT \ln X_{water}$, where X_{water} is the mole fraction of water in the solution. The molar free energy of the solid can be written as $G°_{solid}$. Derive an expression for the change in the melting temperature of ice by equating the free energies in the two phases, differentiating the resulting equation with respect to temperature, integrating from a mole fraction of 1 (pure solvent) to the mole fraction of the solution, and noting that $\ln X_{water} = \ln(1 - X_{urea}) \approx -X_{urea}$. (This relationship is the series expansion of the logarithm for small values of X_{urea}. Since the concentration of water is about 55 M, this is a good approximation.) With the relationship derived, estimate the decrease in the melting temperature of ice for an 1 M urea solution. The heat of fusion of water is 1440 cal/mol.

2-6. What is the maximum amount of work that can be obtained from hydrolyzing 1 mole of ATP to ADP and phosphate at 298 K? Assume that the concentrations of all reactants are 0.01 M and $\Delta G°$ is -32.5 kJ/mol. If the conversion of free energy to mechanical work is 100% efficient, how many moles of ATP would have to be hydrolyzed to lift a 100 kg weight 1 meter high?

CHAPTER 3

Applications of Thermodynamics to Biological Systems

3.1 BIOCHEMICAL REACTIONS

From the discussion in the previous chapter, it should be clear that standard free energy and enthalpy changes can be calculated for chemical reactions from available tables of standard free energies and enthalpies of formation (e.g., Appendix 1). Even if a particular reaction has not been studied, it is frequently possible to combine the standard free energy and enthalpy changes of other known reactions to obtain the standard free energy change of the desired reaction (e.g., Appendix 2). Knowledge of the standard free energy change permits calculation of the equilibrium constant and vice versa. When considering biochemical reactions, the exact nature of the "standard state" can be confusing. Usually the various possible ionization states of the reactants are not specified, so that the total concentration of each species is used in the expression for the equilibrium constant. For example, for ATP, the standard state would normally be l M, but this includes all possible states of protonation, and if magnesium ion is present, the sum of the metal complex and uncomplexed species. Furthermore, pH 7 is usually selected as the standard condition, so that the activity of the hydrogen ion is set equal to 1 at pH 7, and the standard free energy of formation of the hydrogen ion is set equal to zero at pH 7. Therefore, the hydrogen ion is not usually included when writing stoichiometric equations. Finally, the activity of water is set equal to 1 for reactions in dilute solutions. This assumption is justified since the concentration of water is essentially constant. However, a word of caution is needed here as the standard free energy and enthalpy of formation for water must be explicitly included if it is a reactant. Again, the choice of standard states can readily be incorporated into the standard free energy change and standard free energy of formation. The best way to avoid ambiguity is to explicitly write down the chemical reaction and the conditions (pH, salt, etc.) to which a given free energy change refers.

As a specific example, let's again consider the reaction catalyzed by hexokinase, the phosphorylation of glucose by ATP. We have previously calculated the standard enthalpy change for this reaction from the known enthalpies of hydrolysis. This can also be done for the standard free energy change of the hexokinase reaction since the standard free energies of hydrolysis are known. As before, the "standard states" are 1 M, at 1 atmosphere, 298 K, pH 7.0, pMg 3.0, and an ionic strength of 0.25 M.

	$\Delta G°/(\text{kJ/mol})$	$\Delta H°/(\text{kJ/mol})$
$G + P_i \rightleftharpoons G6P + H_2O$	11.6	0.5
$ATP + H_2O \rightleftharpoons ADP + P_i$	−32.5	−30.9
$G + ATP \rightleftharpoons G6P + ADP$	−20.9	−30.4

The equilibrium constant for the hexokinase reaction can be calculated from the standard free energy change:

$$\Delta G° = - RT \ln K \quad \text{or} \quad K = 4630 \ \{=([G6P][ADP])/([G][ATP])\}$$

Since the standard enthalpy change is known, −30.4 kJ/mol, the equilibrium constant can be calculated at other temperatures by use of Eq. 2-45. For example, at 310 K (37°C), the equilibrium constant is 2880. The entropy change for this reaction can be calculated from the relationship $\Delta G° = \Delta H° - T \Delta S°$.

How are standard free energy changes for reactions measured? The obvious answer is by measurement of the equilibrium constant and use of Eq. 2-36. However, in order to determine an equilibrium constant, a measurable amount of both reactants and products must be present. For reactions that essentially go to completion, a sequence of reactions must be found with measurable equilibrium constants that can be summed to total the reaction of interest. For example, the equilibrium constant for the hydrolysis of ATP to ADP and P_i calculated from the above standard free energy change is 5×10^5 M under "standard conditions." Obviously, this constant would be very difficult to measure directly. The following reactions, however, could be used to calculate this equilibrium constant (and the standard free energy change):

$$G + ATP \rightleftharpoons G6P + ADP$$

$$\frac{G6P + H_2O \rightleftharpoons G + P_i}{ATP + H_2O \rightleftharpoons ADP + P_i}$$

The equilibrium constant for the first reaction is about 4600 and for the second reaction is about 110 M, both of which can be measured experimentally. Tables of standard free energies of formation (Appendix 1) and standard free energy changes for reactions (Appendix 2) can be constructed using this methodology.

Knowledge of the equilibrium constants of biological reactions and their temperature dependencies is of great importance for understanding metabolic regulation. It can also be of practical importance in the design of laboratory experiments, for example, in the development of coupled assays.

3.2 METABOLIC CYCLES

Thermodynamics is particularly useful for understanding metabolism. Metabolism consists of many different sets of reactions, each set, or metabolic cycle, designed to utilize and produce very specific molecules. The reactions within a cycle are coupled in that the product of one reaction becomes the reactant for the next reaction in the cycle. These coupled reactions can be very conveniently characterized using thermodynamic concepts. Before considering a specific metabolic cycle, let's consider some general thermodynamic properties of coupled reactions. As a very simple illustration, consider the coupled reactions

$$A \rightarrow B \rightarrow C \rightarrow D$$

The free energy changes for the first three reactions can be written as

$$A \rightarrow B \qquad \Delta G_{AB} = \Delta G^{\circ}_{AB} + RT \ln([B]/[A])$$

$$B \rightarrow C \qquad \Delta G_{BC} = \Delta G^{\circ}_{BC} + RT \ln([C]/[B])$$

$$C \rightarrow D \qquad \Delta G_{CD} = \Delta G^{\circ}_{CD} + RT \ln([D]/[C])$$

The free energy for the overall conversion of A to D can be obtained by adding these free energies:

$$A \rightarrow D \qquad \Delta G_{AD} = \Delta G^{\circ}_{AB} + \Delta G^{\circ}_{BC} + \Delta G^{\circ}_{CD} + RT \ln \{([B][C][D)/([A][B][C])\}$$

$$\Delta G_{AD} = \Delta G^{\circ}_{AD} + RT \ln([D]/[A])$$

Whether A can be converted to D depends on the standard free energies for the three individual reactions and the concentrations of A and D. The concentrations of B and C are of no consequence! Note that since the total standard free energy determines whether A will be converted to D, it is possible for one of the standard free energy changes of the intermediate steps to be very unfavorable (positive) if it is balanced by a very favorable (negative) standard free energy change.

Metabolic pathways contain hundreds of different reactions, each catalyzed by a specific enzyme. Although the thermodynamic analysis shown above indicates that only the initial and final states need be considered, it is useful to analyze a metabolic pathway to see how the individual steps are coupled to each other through the associated free energy changes. The specific metabolic pathway we will examine is anaerobic glycolysis. Anaerobic glycolysis is the sequence of reactions that metabolizes glucose into lactate and also produces ATP, the physiological energy currency. As we have seen, the standard free energy for the hydrolysis of ATP is quite large and negative so that the hydrolysis of ATP can be coupled to reactions with an unfavorable free energy change. The sequence of reactions involved in anaerobic glycolysis is shown in Table 3-1, along with the standard free energy changes. The standard states are as usual for biochemical reactions, namely, pH 7 and 1 M for reactants, with the concentration of each reactant as the sum of all ionized species. The activity of water is assumed to be unity. (This is somewhat different from Appendixes 1 and 2, where ionic strength and, in some cases, the magnesium ion concentration are specified.)

TABLE 3-1. Free Energy Changes for Anaerobic Glycolysis

Reaction	$\Delta G°$ (kJ/mol)	ΔG (kJ/mol)
Part One		
Glucose + ATP $\rightleftharpoons$ Glucose-6-P + ADP	−16.7	−33.3
Glucose-6-P $\rightleftharpoons$ Fructose-6-P	1.7	−2.7
Fructose-6-P + ATP $\rightleftharpoons$ Fructose-1,6-bisphosphate + ADP	−14.2	−18.6
Fructose-1,6-bisphosphate $\rightleftharpoons$ Dihydroxyacetone-P + Glyceraldehyde-3-P	23.9	0.7
Dihydroxyacetone-P $\rightleftharpoons$ Glyceraldehye-3-P	7.5	2.6
Glucose + 2 ATP $\rightleftharpoons$ 2 ADP + 2 Glyceraldehyde-3-P	2.2	−51.3
Part Two		
Glyceraldehyde-3-P + P_i + NAD^+ $\rightleftharpoons$ 1,3-Bisphosphoglycerate + NADH	6.3	−1.0
1,3-Bisphosphoglycerate + ADP $\rightleftharpoons$ 3-P-Glycerate + ATP	−18.9	−0.6
3-Phosphoglycerate $\rightleftharpoons$ 2-Phosphoglycerate	4.4	1.0
2-Phosphoglycerate $\rightleftharpoons$ Phosphoenolpyruvate + H_2O	1.8	1.1
Phosphoenolpyruvate + ADP $\rightleftharpoons$ Pyruvate + ATP	−31.7	−23.3
Pyruvate + NADH $\rightleftharpoons$ Lactate + NAD^+	−25.2	1.9
Glyceraldehyde-3P + P_i + 2 ADP $\rightleftharpoons$ Lactate + 2 ATP + H_2O	−63.3	−20.9

Source: Adapted from R. H. Garrett and C. M. Grisham, *Biochemistry*, Saunders College Publishing, Philadelphia, 1995, pp. 569–597.

The overall reaction for anaerobic glycolysis is

$$\text{Glucose} + 2\ P_i + 2\ \text{ADP} \rightleftharpoons 2\ \text{Lactate} + 2\ \text{ATP} + 2\ H_2O \qquad (3\text{-}1)$$

In anaerobic metabolism, the pyruvate produced in the second to the last step is converted to lactate in muscle during active exercise. In aerobic metabolism, the pyruvate that is produced in the second to the last step is transported to the mitochondria, where it is oxidized to carbon dioxide and water in the citric acid cycle. The reactions in Table 3-1 can be divided into two parts. The first part produces 2 moles of glyceraldehyde-3-phosphate according to the overall reaction

$$\text{Glucose} + 2\ \text{ATP} \rightleftharpoons 2\ \text{ADP} + 2\ \text{Glyceraldehyde-3-phosphate} \qquad (3\text{-}2)$$

Note that this reaction actually requires 2 moles of ATP. However, the second part of the cycle produces 4 moles of ATP with the overall reaction

$$\text{Glyceraldehyde-3-phosphate} + P_i + 2\ \text{ADP} \rightleftharpoons \text{Lactate} + 2\ \text{ATP} + H_2O \qquad (3\text{-}3)$$

Since 2 moles of glyceraldehyde-3-phosphate are produced in part one of the cycle, Eq. 3-3 must be multiplied by two and added to Eq. 3-2 to give the overall reaction, Eq. 3-1. The standard free energies, of course, also must be multiplied by two for the reactions in part two of glycolysis and added to those for the reactions in part one to give the overall standard free energy change, −124.4 kJ/mol. If the standard free energies of formation in Appendix 1 are used to calculate the standard free energy change for Eq. 3-3, a value of −128.6 kJ/mol is obtained. This small difference can be attributed to somewhat different standard states. Unfortunately, not all of the necessary free energies of formation are available in Appendix 1 to calculate standard free energy changes for all of the reactions in Table 3-1.

The standard enthalpy change for Eq. 3-3 is −63.0 kJ/mol, so that a substantial amount of heat is produced by glycolysis. The standard entropy change can be calculated from the known standard free energy and enthalpy changes and is 220 J/(mol·K). Thus, both the standard enthalpy and entropy changes are favorable for the reaction to proceed. It is interesting to compare these numbers with those for the direct oxidation of glucose:

$$\text{Glucose(s)} + 6\,O_2(g) \rightleftharpoons 6\,CO_2(g) + 6\,H_2O(\ell) \tag{3-4}$$

The standard free energy change for this reaction is −2878.4 kJ/mol; the standard enthalpy change is −2801.6 kJ/mol; and the standard entropy change is −259 J/(mol·K). This process produces a very large amount of heat relative to that produced by the metabolic cycle. This would not be very useful for physiological systems.

As we have stressed previously, for a single reaction, it is not the standard free energy change that must be considered in determining whether products are formed, it is the free energy for the particular concentrations of reactants that are present. In order to calculate the free energy changes for the reactions in Table 3-1, the concentrations of metabolites must be known. These concentrations have been determined in erythrocytes and are summarized in Table 3-2. The free energy changes calculated with these concentrations, the standard free energy changes, and Eq. 2-35 are included in Table 3-1. The additional assumptions have been made that these concentrations are valid at 298 K, although they were determined at 310 K, and $[NADH]/[NAD^+] = 1.0 \times 10^{-3}$. (We have elected to carry out calculations at 298 K where the standard free energies are known, rather than at the physiological temperature of 310 K. This does not alter any of the conclusions reached.)

Consideration of free energy changes, rather than standard free energy changes, produces some interesting changes. The overall standard free energy change for the first part of glycolysis is +2.2 kJ/mol, whereas the free energy change is −51.3 kJ/mol. In contrast, the standard free energy change for the second part of glycolysis is much more negative than the free energy change. Thus, when considering the coupling of chemical reactions, considerable care must be exercised in making comparisons. Of course, as we stated at the beginning, the concentrations of the intermediates are of no consequence in determining the overall free energy changes. The reader might wish to confirm that this is indeed the case.

Before we leave our discussion of glycolysis, it is worth addressing the individual reactions in the cycle. All of the reactions are catalyzed by enzymes. If this were not the case, the reactions would occur much too slowly to be physiologically relevant.

TABLE 3-2. Steady-State Concentrations for Glycolytic Intermediates in Erythrocytes

Metabolite	Concentration (mM)
Glucose	5.0
Glucose-6-phosphate	0.083
Fructose-6-phosphate	0.014
Fructose-1,6-bisphosphate	0.031
Dihydroxyacetone phosphate	0.14
Glyceraldehyde-3-phosphate	0.019
1,3-Bisphosphoglycerate	0.001
2,3-Bisphosphoglycerate	4.0
3-Phosphoglycerate	0.12
2-Phosphoglycerate	0.030
Phosphoenolpyruvate	0.023
Pyruvate	0.051
Lactate	2.9
ATP	1.85
ADP	0.14
P_i	1.0

Source: Adapted from S. Minakami and H. Yoshikawa, *Biochem. Biophys. Res. Commun.* **18**, 345 (1965).

The first step is the very favorable phosphorylation of glucose—both the standard free energy change and free energy change are favorable. The advantage to the cell of phosphorylating glucose is that creating a charged molecule prevents it from diffusing out of the cell. Furthermore, the intracellular concentration of glucose is lowered so that if the concentration of glucose is high on the outside of the cell, more glucose will diffuse into the cell. The second step, the isomerization of glucose-6-phosphate to fructose-6-phosphate, has a somewhat unfavorable standard free energy change, but the free energy change is favorable enough for the reaction to proceed. The next step has a very favorable standard free energy change, as well as a favorable free energy change, because the phosphorylation of fructose-6-phosphate is coupled to the hydrolysis of ATP. This very irreversible step is the commitment by the cell to metabolize glucose, rather than to store it. The next two steps produce glyceraldehyde-3-phosphate, the fuel for the second half of glycolysis that produces 4 moles of ATP. Both of these reactions have very unfavorable standard free energy changes, although the free energy changes are only slightly positive. This completes the first part of glycolysis. Note that 2 moles of ATP have been utilized to produce the final product, glyceraldehyde-3-phosphate. As noted previously, the standard free energy change for this first part is actually unfavorable, but the free energy change is quite favorable.

The purpose of the second part of glycolysis is to convert a substantial portion of the metabolic energy of glucose into ATP. In comparing the standard free energy changes with the free energy changes of the individual steps, it is worth noting that the second step, the formation of 3-phosphoglycerate, has a very favorable standard free energy change, yet the free energy change is approximately zero so that this reaction is approximately at equilibrium. This is true of essentially all of the reactions in this part of glycolysis, except for the reaction that produces pyruvate, where both the standard free energy change and the free energy change are quite negative. Both the overall standard free energy change and the free energy change for the second part of glycolysis are quite favorable.

This thermodynamic analysis of glycolysis is a good example of how thermodynamics can provide a framework for understanding the many coupled reactions occurring in biology. It also illustrates how metabolism is utilized to produce molecules such as ATP that can be used to drive other physiological reactions, rather than converting most of the free energy to heat.

3.3 DIRECT SYNTHESIS OF ATP

As we have seen, the standard free energy change for the hydrolysis of ATP to ADP and P_i is -32.5 kJ/mol, so it seems unlikely that ATP would be synthesized by the reverse of this reaction. The concentrations of reactants cannot be adjusted sufficiently to make the overall free energy favorable. Yet, we know that ATP is synthesized directly from ADP and P_i in mitochondria. For many years, people in this field grappled with how this might happen: Both probable and improbable mechanisms were proposed. In 1961 Peter Mitchell proposed that the synthesis of ATP occurred due to a coupling of the chemical reaction with a proton gradient across the membrane (1). This hypothesis was quickly verified by experiments, and he received a Nobel Prize for this work.

The enzyme responsible for ATP synthesis, ATP synthase, consists of a protein "ball," which carries out the catalytic function, coupled to membrane-bound proteins through which protons can be transported. The process of ATP synthesis is shown schematically in Figure 3-1. Although this enzyme is found only in mitochondria in humans, it is quite ubiquitous in nature and is found in chloroplasts, bacteria, and yeast, among others. The chemiosmotic hypothesis states that a pH gradient is established across the membrane by a series of electron transfer reactions, and that ATP synthesis is accompanied by the simultaneous transport of protons across the membrane. The overall reaction can be written as the sum of two reactions:

$$ADP + P_i \rightleftharpoons ATP + H_2O$$

$$n\,H^+_{out} \rightleftharpoons n\,H^+_{in} \qquad (3\text{-}5)$$

$$\overline{ADP + P_i + n\,H^+_{out} \rightleftharpoons ATP + n\,H^+_{in} + H_2O}$$

The value of n has been determined to be 3 (see Ref. 2). The free energy change for the transport of protons in Eq. 3-5 can be written as

$$\Delta G = 3RT \ln([H^+_{in}]/[H^+_{out}]) \qquad (3\text{-}6)$$

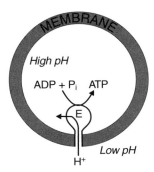

FIGURE 3-1. Schematic representation of ATP synthase, E, in mitochondria. The enzyme structure inside the mitochondria contains the catalytic sites. Protons are pumped from the outside of the membrane to the inside as ATP is synthesized.

(The standard free energy change for this process is zero since the standard state for the hydrogen ion is the same on both sides of the membrane.)

At 298 K, a pH differential of one unit gives a free energy change of −17.1 kJ/mol. The actual physiological situation is even more favorable as a membrane potential exists whereby the membrane is more negative on the inside relative to the outside. If we utilize the extended chemical potential, Eq. 2-49, an additional term is added to Eq. 3-6 equal to 3 $F\Psi$, where 3 is the number of protons transported, F is the Faraday, and Ψ is the membrane potential. For a membrane potential of −100 millivolts, −29 kJ would be added to the free energy change in Eq. 3-6.

The standard free energy for the synthesis of ATP from ADP and P_i is +32.5 kJ/mol, but we need to know the free energy change under physiological conditions. Although the concentrations of the reactants are not known exactly, we can estimate that the ratio of ATP to ADP is about 100, and the concentration of phosphate is 1–10 mM. This makes the ratio [ATP]/([ADP][P_i]) equal to 100–1000. The free energy change at 298 K is

$$\Delta G = 32.5 + RT \ln(100{-}1000) = 32.5 + (11.4{-}17.1) = 43.9{-}49.6 \text{ kJ/mol}$$

Thus, the coupling of the synthesis of ATP to a modest proton gradient and membrane potential can readily provide the necessary free energy for the overall reaction to occur.

The principle to be learned from this example is that the coupling of free energies is very general. It can involve chemical reactions only, as in glycolysis, or it can involve other processes such as ion transport across membranes, as in this example.

3.4 ESTABLISHMENT OF MEMBRANE ION GRADIENTS BY CHEMICAL REACTIONS

In discussing ATP synthesis, we have not specified how the electrochemical gradient is established. This involves a complex sequence of coupled reactions that we shall not discuss here. Instead, we will discuss a case where the free energy associated with the hydrolysis of ATP is used to establish an ion gradient. The process of signal transduction in the nervous system involves the transport of Na^+ and K^+ across the membrane. Neuronal cells accumulate K^+ and have a deficit of Na^+ relative to the external environment. (This is also true for other mammalian cells.) When an electrical signal is transmitted, this imbalance is altered. The imbalance causes a resting membrane potential of about -70 millivolts. This situation is illustrated in Figure 3-2, with some typical ion concentrations.

How is the ion gradient established? This is done by a specific ATPase, the Na^+/K^+ ATPase, that simultaneously pumps ions and hydrolyzes ATP. The process can be written as the sum of two reactions:

$$2\,K^+_{out} + 3\,Na^+_{in} \rightleftharpoons 2\,K^+_{in} + 3\,Na^+_{out} \tag{3-7}$$

$$ATP + H_2O \rightleftharpoons ADP + P_i$$

$$2\,K^+_{out} + 3\,Na^+_{in} + ATP + H_2O \rightleftharpoons 2\,K^+_{in} + 3\,Na^+_{out} + ADP + P_i$$

The stoichiometry has been established by experimental measurements. The free energy change for the first step is

$$\Delta G = RT\ln\{([Na^+_{out}]^3[K^+_{in}]^2)/([Na^+_{in}]^3[K^+_{out}]^2)\} + 3\,F\Psi - 2\,F\Psi$$

With $\Psi = 70$ millivolts, $T = 298$ K, and the concentrations of Na^+ and K^+ in Figure 3-2, $\Delta G = 41.3$ kJ/mol. Note that the membrane potential produces a favorable (negative) free energy change for the transport of K^+ and an unfavorable free energy change (positive) for the transport of Na^+. We have previously calculated that ΔG for ATP hydrolysis is $-(43.9 - 49.6)$kJ/mol. Therefore, the hydrolysis of ATP is sufficient to es-

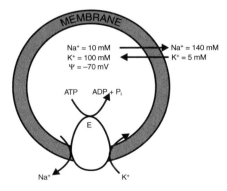

FIGURE 3-2. Schematic representation of the Na^+ and K^+ gradients in a cell. The free energy necessary for the creation of these gradients is generated by the enzymatic hydrolysis of ATP. The Na^+/K^+ ATPase is designated as E, and typical concentrations of the ions and membrane potential, Ψ, are given.

tablish the ion gradient and accompanying membrane potential. This process is called *active transport*.

The coupling of free energies in biological systems can be used to understand many of the events occurring, and many other interesting examples exist. However, we will now turn to consideration of protein and nucleic acid structures in terms of thermodynamics.

3.5 PROTEIN STRUCTURE

An extensive discussion of protein structure is beyond the scope of this treatise, but it is illuminating to discuss a few aspects of proteins in terms of thermodynamics. Many excellent discussions of protein structure are available (cf. Refs. 3–5). Proteins, of course, are polymers of amino acids, and the amino acids contain both polar and non-polar groups. The structures of the 20 common amino acids are given in Appendix 3, and the covalent structure of a polypeptide chain is shown in Figure 3-3. We will first consider the role of nonpolar groups in proteins, that is, amino acid side chains containing methylene and methyl groups and aromatic residues.

As a starting point, let us consider the thermodynamics of transferring the hydrocarbons methane and ethane from an organic solvent to water. The thermodynamic parameters can be measured for this process simply by determining the solubility of the hydrocarbons in each of the solvents. Since the pure hydrocarbon is in equilibrium with each of the saturated solutions, the two saturated solutions must be in equilibrium with each other. The transfer reaction can be written as the sum of the solubility equilibria:

$$\text{Hydrocarbon (organic solvent)} \rightleftharpoons \text{Hydrocarbon} \qquad (3\text{-}8)$$

$$\text{Hydrocarbon} \rightleftharpoons \text{Hydrocarbon (water)}$$

$$\text{Hydrocarbon (organic solvent)} \rightleftharpoons \text{Hydrocarbon (water)}$$

The transfer free energy is

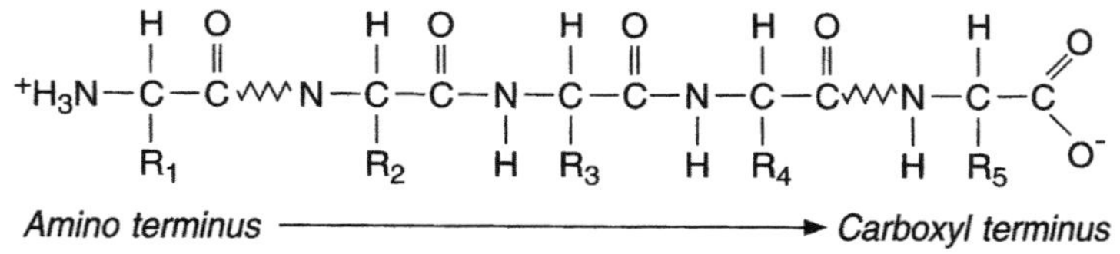

FIGURE 3-3. A polypeptide chain. The amino acid side chains are represented by R_i, and the amino and carboxyl termini are shown.

$$\Delta G_t = RT \ln([X_O]/[X_W])$$

where X_O and X_W are the mole fraction solubilities in the organic solvent and water, respectively. (The use of mole fractions is the appropriate concentration scale in this instance because of standard state considerations that we will not deal with here.) Measurement of the temperature dependence of the transfer free energy will permit ΔH and ΔS to be determined (Eqs. 2-42 and 2-43). The thermodynamic parameters obtained at 298 K in several organic solvents are summarized in Table 3-3.

As expected, the free energy change is unfavorable—hydrocarbons do not dissolve readily in water. However, the energy change is favorable so that the negative entropy change is responsible for the unfavorable free energy. What is the reason for the negative entropy change? It is due to the fact that the normal water structure is broken by the insertion of the hydrocarbon, and the water molecules tend to form a hydrogen-bonded structure around the hydrocarbon that is much more ordered than the normal structure of water. Thus, it is the formation of ice-like ordered structures around the hydrocarbon that is the source of the negative entropy change.

To make this more relevant to proteins, the free energy of transfer for amino acids from a hydrocarbon-like environment (ethanol) to water has been measured. This is done by determining the solubility of the amino acids in ethanol and water. Similar to before, in both cases the amino acid in solution is in equilibrium with the solid, so that the transfer free energy can be measured:

$$\text{Amino acid (ethanol)} \rightleftharpoons \text{Amino acid (solid)} \qquad (3\text{-}9)$$

$$\text{Amino acid (solid)} \rightleftharpoons \text{Amino acid (water)}$$

$$\text{Amino acid (ethanol)} \rightleftharpoons \text{Amino acid (water)}$$

Some of the results obtained are summarized in Table 3-4. In order to interpret these results, we must remember that in ethanol the amino acid is uncharged (NH_2—RCH—COOH), whereas in water it is a zwitterion (NH_3^+—RCH—COO$^-$). If glycine,

TABLE 3-3. Thermodynamic Parameters for Hydrocarbon Transfer from Organic Solvents to Water at 298 K

Transfer Reaction	ΔS_t [cal/(mol·K)]	ΔH_t (kcal/mol)	ΔG_t (kcal/mol)
CH_4, benzene to water	−18	−2.8	+2.6
CH_4, ether to water	−19	−2.4	+3.3
CH_4, CCl_4 to water	−18	−2.5	+2.9
C_2H_6, benzene to water	−20	−2.2	+3.8
C_2H_6, CCl_4 to water	−18	−1.7	+3.7

Source: Adapted from W. Kauzmann, *Adv. Protein Chem.* **14**, 1 (1959).

TABLE 3-4. Free Energy Changes for Transferring Amino Acids from Ethanol to Water at 298 K

Compound	ΔG_t (kcal/mol)	$\Delta G_{t,\text{side chain}}$ (kcal/mol)
Glycine	−4.63	—
Alanine	−3.90	+0.73
Valine	−2.94	+1.69
Leucine	−2.21	+2.42
Isoleucine	−1.69	+2.97
Phenylalanine	−1.98	+2.60
Tyrosine	−1.78	+2.85

Source: Adapted from C. Tanford, *J. Am. Chem. Soc.* **84**, 4240 (1962).

which has no side chains, is taken as the standard, then subtracting its transfer free energy from the transfer free energies of the other amino acids will give the transfer free energy for the amino acid side chains. The overall transfer free energy for all of the amino acids is negative (favorable) because the charged amino and carboxyl groups are solvated by water, a highly favorable interaction. However, the standard free energy changes of transfer for the hydrophobic side chains are all positive, as was seen for the transfer of methane and ethane from an organic solvent into water.

What is the relevance of such data for protein structure? These data indicate that the apolar side chains of amino acid side chains would prefer to be in a nonaqueous environment. That is, they prefer to cluster together. This is, in fact, true for most proteins. The hydrophobic groups aggregate together on the interior of the protein, forming a hydrophobic core, with the more polar groups tending to be on the outside interacting with water. This important concept was enunciated by Walter Kauzmann in 1959 (6). Although the interactions associated with forming the hydrophobic core are often called hydrophobic bonds, it is important to remember that the driving force for forming a hydrophobic core is not the direct interactions between hydrophobic groups; instead, it is the release of water molecules from the ice-like structures that surround hydrophobic groups in water. The thermodynamic analysis clearly indicates that the formation of "hydrophobic bonds" is an entropy-driven process. In fact, ΔH_t for the transfer of methane and ethane from water to organic solvent is positive, that is, energetically unfavorable. This means that an increase in temperature will tend to strengthen the hydrophobic bonding—if ΔH remains positive over the temperature range under consideration.

The formation of hydrogen bonds in protein structures is prevalent, and we will now consider the thermodynamics of hydrogen bond formation. Thermodynamic studies have been made of many different types of hydrogen bonds. A few selected examples are summarized in Table 3-5. The first is the formation of a hydrogen-bonded dimer of acetic acid molecules in the gas phase:

$$2CH_3-\underset{O-H}{\overset{O}{C}} \; \text{(g)} \rightleftharpoons CH_3-C\underset{O-H----O}{\overset{O----H-O}{\diagup}}C-CH_3 \text{(g)} \qquad (3\text{-}10)$$

As expected, both $\Delta H°$ and $\Delta S°$ are negative. The former represents the energy produced in forming two hydrogen bonds, and the latter is due to two molecules forming a dimer (a more ordered system). Since two hydrogen bonds are formed the enthalpy change associated with a single hydrogen bond is about -7 kcal/mol.

The second example is the dimerization of N-methylacetamide, a good model for hydrogen bonding involving the peptide bond:

$$2CH_3-\underset{H}{N}-\overset{O}{\overset{\|}{C}}-CH_3 \rightleftharpoons \begin{array}{cc} CH_3 & CH_3 \\ C=O-----HN \\ HN & C=O \\ CH_3 & CH_3 \end{array} \qquad (3\text{-}11)$$

This dimerization was studied in a variety of solvents, and the results in carbon tetrachloride, dioxane, and water are included in Table 3-5. In carbon tetrachloride, the solvent does not compete for the hydrogen bonds of N-methylacetamide, and the results are not very different from those for acetic acid dimerization in the gas phase, namely, $\Delta H°$ for formation of a single hydrogen bond is about -4 kcal/mol and $\Delta S°$ is about -11 cal/(mol·K). However, in water, which can also form hydrogen bonds with N-methylacetamide and is present at a concentration of about 55 M, $\Delta H°$ is approximately zero. This indicates that there is not a significant enthalpy difference between the water–N-methylacetamide hydrogen bond and the N-methylacetamide–N-methylacetamide hydrogen bond. Note that the standard free energy change in water is 3 kcal/mol, so that the amount of dimer formed is very small even at 1 M N-methyl-

TABLE 3-5. Thermodynamic Parameters for Hydrogen Bond Formation[a]

Reactant	Solvent	$\Delta G°_{298}$ (kcal/mol)	$\Delta H°_{298}$ (kcal/mol)	$\Delta S°_{298}$ [cal/(mol·k)]
CH_3COOH[b]	Gas	-8.02	-15.9	-26.6
CH_3CONH_2[c]	CCl_4	-0.92	-4.2	-11
	Dioxane	0.39	-0.8	-4
	H_2O	3.1	0.0	-10

[a]Standard state is 1 M.
[b]See J. O. Halford, *J. Chem. Phys.* **9**, 859 (1941).
[c]See I. M. Klotz and J. S. Franzen, *J. Am. Chem. Soc.* **84**, 3461 (1962).

acetamide. Dioxane has two oxygens that can accept a hydrogen bond, so that $\Delta H°$ is only slightly negative.

The conclusion reached from these data (and considerably more data not presented) is that the stability of a water–protein hydrogen bond is similar to that of intramolecular protein hydrogen bonds. Therefore, a single intramolecular hydrogen bond on the surface of the protein is unlikely to be a strong stabilizing factor. On the other hand, if the hydrophobic interior of a protein excludes water, a strong hydrogen bond might exist in the interior of a protein. This statement is complicated by the fact that proteins have "breathing" motions; that is, the protein structure continually opens and closes with very small motions so that completely excluding water from the interior may be difficult.

The role of the hydrogen bond in stabilizing proteins is still a matter of some debate. The statement is often made that hydrophobic interactions are the primary source of protein stability and hydrogen bonding provides specificity (6). There is no doubt that extended hydrogen-bonded systems are extremely important structural elements in proteins. Two examples are given in Color Plates I and II, the α-helix and the β-pleated sheet. The α-helix is a spiral structure with 3.6 amino acids per turn of the helix. Every peptide carbonyl is a hydrogen bond acceptor for the peptide N—H four residues away. This structure is found in many different proteins. The β-pleated sheet is also a prevalent structure in proteins. In the β-pleated sheet, each chain is a "pleated" helix. Again, all of the peptide bonds participate in hydrogen bonding, but the bonds are all between chains, rather than intrachain.

Finally, we will say a few words about the role of electrostatic interactions in protein structures. The discussion will be confined to charge–charge interactions, even though more subtle interactions involving, for example, dipoles and/or induced dipoles are important. Many of the amino acids have side chains with ionizable groups, so that a protein contains many acids and bases. For example, a carboxyl group can ionize according to the scheme

$$P\text{-COOH} \rightleftharpoons P\text{-COO}^- + H^+ \tag{3-12}$$

Here P designates the protein. Although not shown, water plays an extremely important role in this equilibrium. Water is a strong dipole and strongly solvates ions, forming a hydration "sheath." Acetic acid is a reasonable model for this reaction: The thermodynamic parameters characterizing its ionization are $\Delta G° = 6.6$ kcal/mol, $\Delta H° = 0$, and $\Delta S° = -22$ cal/(mol·K). The negative entropy change is due to the ordering of water molecules around the ions. Note that the ionization process is thermally neutral. The enthalpy changes associated with the solvation of ions are generally negative but, in this case, are balanced by the enthalpic change associated with breaking the oxygen–hydrogen bond.

The ionization constants for ionizable groups on proteins are generally not the same as those of simple model compounds. This is because the ionization process is influenced by the charge of the protein created by other ionizable groups and, in some cases, by special structural features of the protein. This factor can be included explicitly in a thermodynamic analysis of protein ionizations by writing the free energy associated with ionization as the sum of the free energy for the model compound, or the intrinsic free energy, and the free energy of interaction:

$$\Delta G_{ionization} = \Delta G^{\circ}_{intrinsic} + \Delta G_{interaction} \qquad (3\text{-}13)$$

The ionization properties of proteins have been studied extensively, both experimentally and theoretically. For our purposes, it is important to recognize that the strong solvation of ions means that charged groups will tend to be on the outside of the protein, readily accessible to water.

A reasonable question to ask is: If there are so many charged groups on proteins, won't they influence the structure simply because charged groups of opposite sign attract, and those with the same sign repel? This is certainly the case—at very high pH a protein becomes very negatively charged and the interactions between negative charges can eventually cause the native structure to disappear. Similarly, at very low pH, the interactions between positive charges can cause disruption of the native structure. Conversely, the formation of a salt linkage between groups with opposite charges can stabilize structures. We can make a very simple thermodynamic analysis of charge–charge interactions by recognizing that the free energy of formation of an ion pair is simply the work necessary to bring the ions to within a specified distance, a:

$$\Delta G = z_1 z_2 e^2 / D a \qquad (3\text{-}14)$$

Here the z_i are the ionic valences, e is the charge on an electron, and D is the dielectric constant. The assumption of a simple coulombic potential is a gross oversimplification but suffices for our purposes. A more complete potential would include the ionic environment of the medium, the structure of the ions, and the microscopic structure of the solvent. If the valences are assumed to be 1, the distance of closest approach 4 Å, and the dielectric constant of water 80, ΔG is about 1 kcal/mol. Thus, the interaction energy is not very large in water. However, the interior of a protein is more like an organic solvent, and organic solvents have dielectric constants of about 3, which would significantly increase the free energy of interaction. In fact, salt linkages are rarely found near the surface of proteins but are found in the interior of proteins, as expected.

If the free energy of ionic interactions is given by Eq. 3-14, the enthalpy and entropy can easily be calculated from Eqs. 2-42 and 2-43. Since only the dielectric constant in Eq. 3-14 is temperature dependent, this gives

$$\Delta H = -T^2[d(\Delta G/T)/dT] = (\Delta G/D)[d(DT)/dT] \qquad (3\text{-}15)$$

$$\Delta S = -d\,\Delta G/dT = (\Delta G/D)(dD/dT) \qquad (3\text{-}16)$$

In water, both of these derivatives are negative so that if ΔG is negative, both the entropy and enthalpy changes are positive. The positive entropy change can be rationalized as being due to the release of water of hydration of the ions when the ion pair is formed. The enthalpy change is a balance between the negative enthalpy change from bringing the charges closer and the positive enthalpy change associated with removing the hydration shell. Since the enthalpy change is positive, the strength of the ion pair interaction will increase as the temperature is increased—exactly as for hydrophobic interactions.

Although this is an abbreviated discussion, it is clear that we know a great deal about the thermodynamics of interactions that occur in proteins. Because of this, you might think that we could examine the amino acid sequence of any protein and predict its structure by looking at the possible interactions that occur and finding the structure that has the minimum free energy. This has been a long-standing goal of protein chemistry, but we are not yet able to predict protein structures. Why is this the case? The difficulty is that there are thousands of possible hydrogen bonding interactions, ionic interactions, hydrophobic interactions, etc. As we have seen, each of the individual interactions is associated with small free energy and enthalpy changes—in some cases we cannot even determine the sign. The sum of the positive free energies of interactions is very large, as is the sum of the negative free energies. It is the difference between the positive and negative free energies that determines the structure. So we have two problems: accurately assessing the free energies of individual interactions and then taking the difference between two large numbers to determine which potential structure has the minimum free energy. These are formidable problems, but significant progress has been made toward achieving the ultimate goal of predicting protein structures. The possibility also exists that the structure having the lowest free energy is not the biologically relevant structure. This may be true in a few cases but is unlikely to be a problem for most proteins.

3.6 PROTEIN FOLDING

The folding of proteins into their biologically active structures has obvious physiological importance. In addition, understanding the process in molecular detail is linked directly to understanding protein structure. The study of protein folding, and the reverse process of unfolding, is a major field of research, and we will only explore a few facets of this fascinating subject. For our discussion, we will concentrate on protein unfolding as this is most easily experimentally accessible. There are many ways of unfolding proteins. When the temperature increases, proteins will eventually unfold. From a thermodynamic standpoint, this is because the $T\Delta S$ term eventually dominates in determining the free energy change, and we know that the unfolded state is more disordered than the native state at sufficiently high temperatures.

Chemical denaturants such as acid, base, urea, and guanidine chloride are also often used to unfold proteins. The role of a neutrally charged denaturant such as urea can be understood in thermodynamic terms by considering the free energy of transfer of amino acids from water to urea, again using glycine as a reference. Some representative free energies of transfer for hydrophobic side chains are given in Table 3-6. Note that the free energies are all negative, so that removing hydrophobic side chains from the interior of the protein into 8 M urea is a favorable process.

Protein unfolding can be monitored by many different methods. Probably the most common is circular dichroism in the ultraviolet, which is quite different for native and unfolded structures. Many other spectral and physical methods work equally well. A

TABLE 3-6. Transfer Free Energies for Selected Amino Acids from Water to 8 M Urea at 298 K

Amino Acid	ΔG_t (kcal/mol)	$\Delta G_{t,\text{side chain}}$ (kcal/mol)
Glycine	+0.10	0.0
Alanine	+0.03	−0.07
Leucine	−0.28	−0.38
Phenylalanine	−0.60	−0.70
Tyrosine	−0.63	−0.73

Source: Adapted from P. L. Whitney and C. Tanford, *J. Biol. Chem.* **237**, 1735 (1962).

representative plot of the fraction of denatured protein, f_D, versus temperature is shown in Figure 3-4 for the N-terminal region (amino acid residues 6–85) of λ repressor, a protein from λ phage that binds to DNA and regulates transcription. Results are shown for both no urea and 2 M urea. As expected, the protein is less stable in the presence of 2 M urea. The fraction of protein denatured can be written as

$$f_D = [D]/([D] + [N]) = K/(1 + K) \tag{3-17}$$

where D is the concentration of denatured species, N is the concentration of native species, and K ($= [D]/[N]$) is the equilibrium constant for denaturation. This assumes that the unfolding process can be characterized by only two states—native and denatured. For some proteins, intermediates are formed as the protein unfolds, and a more complicated analysis must be used.

Since the equilibrium constant can be calculated at any point on the curve, and its temperature dependence can be measured, the thermodynamic parameters characterizing the unfolding can be determined. The thermodynamics of unfolding can also

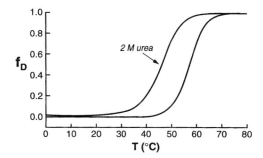

FIGURE 3-4. Fraction of denatured protein, f_D, for λ_{6-85} phage repressor protein as a function of temperature, T, in aqueous solution and in 2 M urea. Data from G. S. Huang and T. G. Oas, *Biochemistry* **35**, 6173 (1996).

conveniently be studied by scanning calorimetry. It is often found that a plot of the logarithm of the equilibrium constant versus $1/T$ is not linear, as predicted, namely,

$$d \ln K/d(1/T) = -\Delta H/R \tag{3-18}$$

This means that the enthalpy change is temperature dependent, or in other words, there is a large heat capacity difference, ΔC_P, between the native and unfolded states.

$$d \Delta H/dT = \Delta C_P \tag{3-19}$$

A typical experimental result for the variation of the equilibrium constant with temperature for λ repressor is given in Figure 3-5a. The plot of $\ln K$ versus $1/T$ goes through a minimum: At higher temperatures, ΔH is positive, whereas at lower temperatures, it is negative. For typical chemical reactions, this plot is a straight line! In Figure 3-5b, the standard free energy change is plotted versus the temperature. The protein is most stable at the maximum in the free energy curve, about 15°C. In Figure 3-5c, $\Delta H°$ and $T \Delta S°$ are plotted versus T. Note that both $\Delta H°$ and $\Delta S°$ are zero at a specific temperature. The practical implication of the positive value of ΔC_P is that the protein unfolds at both high temperatures and low temperatures. This unexpected result is not confined to λ repressor. Most proteins will unfold (denature) at low temperatures, but in many cases the temperatures where cold denaturation is predicted to occur are below 0°C.

Some thermodynamic parameters characterizing protein denaturation are given in Table 3-7 for a few proteins. In all cases, a large positive value of ΔC_P is observed. Note that at room temperature and above, $\Delta H°$ is typically large and positive, as is $\Delta S°$, so that unfolding is an entropically driven process. Before we leave our discussion of

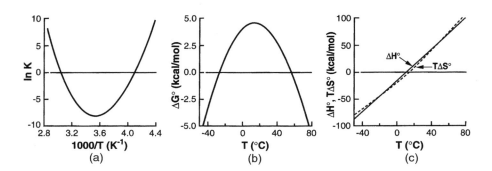

FIGURE 3-5. (a) Plot of the natural logarithm of the equilibrium constant, K, for the denaturation of λ_{6-85} phage repressor protein in aqueous solution versus the reciprocal temperature on the Kelvin scale, $1/T$. (b) Plot of $\Delta G°$ for the denaturation of λ_{6-85} phage repressor protein versus the temperature, T. (c) Plots of $\Delta H°$ and $T \Delta S°$ versus the temperature, T, for the denaturation of λ_{6-85} phage repressor protein. Data from G. S. Huang and T. G. Oas, *Biochemistry* **35**, 6173 (1996).

TABLE 3-7. Representative Thermodynamic Parameters for Thermal Protein Denaturation in Aqueous Solution at 298 K

Protein	$\Delta G°$ (kJ/mol)	$\Delta H°$ (kJ/mol)	$\Delta S°$ [J/(mol·K)]	ΔC_P [kJ/(mol·K)]
Barnase	48.9	307	866	6.9
Chymotrypsin	45.7	268	746	14.1
Cytochrome-c	37.1	89	174	6.8
Lysozyme	57.8	242	618	9.1
Ribonuclease A	27.0	294	896	5.2
λ Repressor$_{6\text{-}85}$[a]	17.7	90.4	244	6.0

Source: Adapted from G. I. Makhatadze and P. L. Privalov, Adv. Prot. Chem. **47**, 307 (1995).
[a] See G. S. Huang and T. G. Oas, Biochemistry **35**, 6173 (1996).

protein folding/unfolding, one more important point should be mentioned. The transition from folded to unfolded states usually occurs over a fairly small change in temperature or denaturant concentration. This is because folding/unfolding is a highly *cooperative* process—once it starts, it proceeds with very small changes in temperature or denaturant. Cooperative processes are quite prevalent in biological systems, particularly when regulation of the process is desired; cooperative processes will be discussed again in Chapter 6.

3.7 NUCLEIC ACID STRUCTURES

We will now briefly consider the structure of nucleic acids in terms of thermodynamics. Again, only a few examples will be considered. Many more complete discussions are available (3,4,7). Let us start with the well-known structure of DNA. Most DNAs consist of two chains that form a double helix. Each chain is a polymer of nucleosides linked by phosphodiester bonds as shown in Figure 3-6. Each nucleoside contains a 2′-deoxyribose sugar and a base, almost always adenine (A), thymine (T), cytosine (C), or guanine (G). The phosphodiester linkage is through the 5′ and 3′ positions on the sugars. By convention, a DNA chain is usually written so that the 5′ end of the molecule is on the left and the 3′ on the right. The B form of the DNA double helix is shown in Color Plate III. The chains are arranged in an antiparallel fashion and form a right-handed helix. The bases are paired through hydrogen bonds on the inside of the double helix, and as might be expected the negatively charged phosphodiester is on the outside. Note the similarity to protein structure as the more hydrophobic groups tend to be on the inside and the hydrated polar groups on the outside. However, DNA is not globular, unlike most proteins, and forms a rod-like structure in isolation. These rods can bend and twist to form very compact structures in cells.

Let us first examine the hydrogen-bonded pairs in DNA. Early in the history of DNA, Erwin Chargaff noted that the fraction of bases that were A was approximately

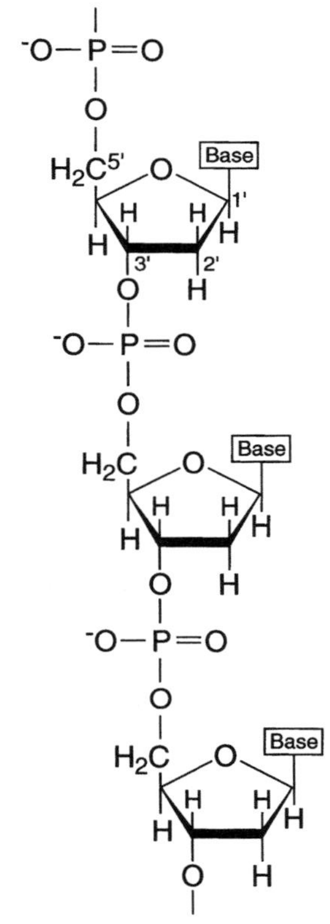

FIGURE 3-6. Structural formula of part of a DNA/RNA chain. In DNA, the bases are usually A, T, C, and G. In RNA, the T is replaced by U, and the 2'-H below the plane of the sugar is replaced by OH.

equal to the fraction that were T in many different DNAs. Similarly, the fraction of bases that were G was equal to the fraction that were C. This finding was important in the postulation of the double helix structure by James Watson and Francis Crick. As shown in Figure 3-7, these bases form hydrogen-bonded pairs, with the A–T pair forming two hydrogen bonds and the G–C pair forming three hydrogen bonds. These are not the only possible hydrogen-bonded pairs that can be formed between the four bases. Table 3-8 gives the thermodynamic parameters associated with the formation of various hydrogen-bonded pairs in deuterochloroform. Instead of thymine, uracil has been used—this is the base normally found in ribonucleic acids and differs only by a methyl group in the 5 position from thymine. It is clear that the A–U pair is favored over the A–A and U–U pairs, and that G–C is favored over G–G and C–C although the data are not very precise. There is no simple explanation for this strong preference—it must be due to more than hydrogen bonding, perhaps an electronic effect within those specific pairs. Furthermore, the preference for A–T in the former three pairs is due to a more favorable enthalpy change. Based on this observation, the

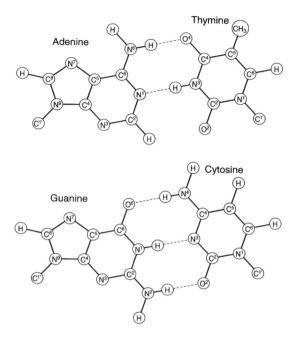

FIGURE 3-7. Formation of the hydrogen-bonded base pairs, A–T and G–C, in DNA. In RNA, U is substituted for T. All of the base rings are aromatic.

entropy change for the latter three pairs has been assumed to be the same in order that $\Delta H°$ can be calculated.

In water solution, the hydrogen bonding between bases is considerably weaker than in organic solvents because of the competition for hydrogen bond formation with water. The assumption is that the base pair hydrogen bonds are sufficiently shielded

TABLE 3-8. Thermodynamic Parameters for Base Pairs in Deuterochloroform at 298 K

Base Pair	$\Delta G°$ (kcal/mol)	$\Delta H°$ (kcal/mol)[a]	$\Delta S°$ [cal/(mol·k)]
A–A	−0.67	−4.0	−11.4
U–U	−1.07	−4.3	−11.0
A–U	−2.72	−6.2	−11.8
C–C	−1.97	−6.3	−15
G–G	−4.1 to −5.4	(−8.5 to −10)	(−15)
G–C	−5.4 to −6.8	(−10.0 to −11.5)	(−15)

Source: Adapted from C. R. Cantor and P. R. Schimmel, *Biophysical Chemistry*, W. H. Freeman, New York, 1980, p. 325. *Data source*: Y. Kyogoku, R. C. Lord, and A. Rich, *Biochim. Biophys. Acta* **179**, 10 (1969).
[a] The $\Delta H°$ values in parentheses have been calculated assuming $\Delta S° = -15$ cal/(mol·K).

TABLE 3-9. Association Constants and Standard Free Energy Changes for Base Stacking in H_2O at 298 K

Base Stack	K (molal^{-1})	$\Delta G°$ (kcal/mol)
A–A	12	−1.50
T–T	0.91	+0.06
C–C	0.91	+0.06
T–C	0.91	+0.06
A–T	3 to 6	−0.70 to −1.10
A–C	3 to 6	−0.70 to −1.10
G–C	4 to 8	−0.80 to −1.20

Source: Adapted from T. N. Solie and J. A. Schellman, *J. Mol. Biol.* **33**, 6 (1968).

from water so that the structure is very stable. However, we know that this is not entirely correct as some "breathing" motions occur, so some exposure to water must also occur. Factors other than hydrogen bonding must contribute to the stability of the double helix.

A consideration comes into play in nucleic acid structures that is not a major concern in proteins, namely, the interactions between the aromatic rings of the bases. The planes of the bases lie over one another so that the π electrons interact; that is, the rings are attracted to each other. This "stacking" reaction can be studied in model systems by measuring the equilibrium constant for the interaction of nucleosides in water, where the hydrogen bonding between nucleosides is negligible. Some representative equilibrium constants for dimer formation are given in Table 3-9. The interaction is quite weak, although it appears that purine–purine interactions are stronger than purine–pyrimidine interactions, which in turn are stronger than pyrimidine–pyrimidine interactions. This is the order expected for π electron interactions. The enthalpy change associated with these interactions is somewhat uncertain but is definitely negative. A value of −3.4 kcal/mol has been obtained for formation of an A–A stack (8).

Is the "stacking" interaction entirely due to π electron interactions, or is the solvent involved, as for hydrophobic interactions in proteins? The "stacking" interaction has been found to be solvent dependent, with water being the most favorable solvent. This suggests that hydrophobic interactions may be involved. However, you should recall that hydrophobic interactions are endothermic, and entropically driven by the release of ordered water molecules around the noninteracting hydrophobic groups. The most likely possibility is that both π electron interactions and hydrophobic effects play a role in base stacking.

3.8 DNA MELTING

One of the reasons for wanting to understand the interactions in the DNA molecule is to understand the stability of DNA since it must come apart and go together as cells reproduce. One way to assess the stability of DNA is to determine its thermal stability. This can readily be studied experimentally because the ultraviolet spectra of stacked bases differ significantly from those of unstacked bases. Thus, if the temperature is raised, the spectrum of DNA changes as it "melts." This is shown schematically in Figure 3-8. Because of the relative stability of the hydrogen bonding, the A–T regions melt at a lower temperature than the G–C regions. The melting of AT and GC polymers can be measured, and as shown in Figure 3-8 the AT structure melts at a lower temperature than a typical DNA, and the GC structure at a higher temperature. The temperature at which half of the DNA structure has disappeared is the melting temperature, T_m. If the transition from fully helical DNA to separated fragments is assumed to be a two-state system, then thermodynamic parameters can be calculated, exactly as for protein unfolding. Thus, the equilibrium constant is 1 ($\Delta G° = 0$) at T_m. Determination of the temperature dependence of the equilibrium constant permits calculation of the standard enthalpy and free energy changes for the "melting" process.

The real situation is more complex than shown, as DNA may not melt in a well-defined single step, and mixtures of homopolynucleotides can form structures more complex than dimers, but a more detailed consideration of these points is beyond the scope of this text. However, it should be mentioned that thermal melting of single-stranded homopolymers is a useful tool for studying stacking interactions: As with DNA, the spectrum will change as the bases unstack. The thermodynamic interpretation of these results is complex (cf. Ref. 3).

From a practical standpoint, many of the above difficulties can be circumvented by the use of model systems to determine the thermodynamics of adding a base pair to a chain of nucleotides. Thus, for example, an oligonucleotide A_n could be added to T_n,

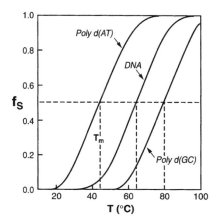

FIGURE 3-8. Hypothetical melting curves for double helix structures of poly d(AT), DNA(containing A–T and G–C base pairs), and poly d(GC). The fraction of chains not in the double helix structure, f_S, is plotted versus the temperature. The melting temperatures, T_m, are indicated by the dashed lines. The shape of the curves and the T_m values are dependent on the length of the chains and their concentrations. This drawing assumes that the concentrations and chain lengths are comparable in all three cases.

and a complex of A_nT_n formed. The same experiment can be done with A_{n+1} and T_{n+1}. If $\Delta G°$ is determined for the formation of each of the complexes, then the difference between the two standard free energies of formation gives the standard free energy change for the formation of one A–T pair. The temperature dependence of this free energy difference can be used to obtain the standard enthalpy and entropy changes for formation of a single A–T pair. With knowledge of the thermodynamic parameters for the formation of single A–T and G–C pairs within a chain, the question can be asked: Does this permit calculation of the thermodynamic stability of a DNA of a given composition? Regrettably, this is not the case. The stability of a given DNA cannot be predicted simply by knowing the fraction of G–C pairs in the DNA.

Remarkably, if the effect of the nearest neighbors of each base pair is taken into account, a reasonable estimate of the thermodynamic stability of DNA can be obtained. This was discovered by studying many different complementary strands of short DNAs and looking for regularities (9,10). The rationale for this procedure lies in the importance of hydrogen bonding between each pair and the stacking interactions with its nearest neighbors. Ten nearest-neighbor parameters suffice for determining the thermodynamic stability of a given DNA sequence, with one additional assumption, namely, the thermodynamic parameters for initiating DNA structure are different from those for adding hydrogen-bonded pairs to an existing chain. This takes into account that getting the two chains together and forming the first base pair is more difficult than adding base pairs to the double helix. Moreover, the G–C hydrogen-bonded pair nucleates the double helix formation better than an A–T pair since it is more stable. In thermodynamic terms, this can be written as

$$\Delta G° = \Delta G°(\text{initiation}) + \Sigma \Delta G°(\text{nearest neighbors})$$

(3-20)

where the sum is over all pairs of nearest neighbor interactions. The thermodynamic parameters necessary to calculate the nearest-neighbor interactions are given in Table 3-10, along with those characterizing the double helix initiation. The best fit of the data on model systems utilizes two additional parameters that are usually only small corrections; this refinement will not be considered here (cf. Ref. 9).

A simple example will illustrate how these data can be used. Consider the reaction

$$\begin{array}{c} \text{5'-A-G-C-T-G-3'} \\ + \\ \text{5'-C-A-G-C-T-3'} \end{array} \quad \rightleftharpoons \quad \begin{array}{c} \text{5'-A-G-C-T-G-3'} \\ \text{3'-T-C-G-A-C-5'} \end{array}$$

The sum of the nearest-neighbor standard free energy changes associated with the formation of the five base pair DNA from Table 3-10 is

TABLE 3-10. Thermodynamic Parameters for Determination of DNA Stability[a]

DNA Pair	$\Delta G°$ (kcal/mol)	$\Delta H°$ (kcal/mol)	$\Delta S°$ [cal/(mol·K)]
5'-A-A 3'-T-T	−1.4	−8.4	−23.6
5'-A-T 3'-T-A	−0.9	−6.5	−18.8
5'-T-A 3'-A-T	−0.8	−6.3	−18.5
5'-A-C 3'-T-G	−1.8	−8.6	−23.0
5'-C-A 3'-G-T	−1.6	−7.4	−19.3
5'-A-G 3'-T-C	−1.3	−6.1	−16.1
5'-G-A 3'-C-T	−1.7	−7.7	−20.3
5'-C-G 3'-G-C	−2.5	−10.1	−25.5
5'-G-C 3'-C-G	−2.5	−11.1	−28.4
5'-C-C 3'-G-G	−2.1	−6.7	−15.6
Initiation, one or more GC hydrogen bond pairs	+1.8	0.0	−5.9
Initiation, no GC hydrogen bond pairs	+2.7	0.0	−9.0

Source: Adapted from J. SantaLucia Jr., H. T. Allawi, and P. A. Seneviratne, *Biochemistry* **35**, 3555 (1996).
[a]pH 7.0, 1 M NaCl, 298 K.

$$\sum \Delta G° = \text{5'-A-G} + \text{5'-G-C} + \text{5'-C-T} + \text{5'-T-G}$$
$$\text{T-C-5'} \quad \text{C-G-5'} \quad \text{G-A-5'} \quad \text{A-C-5'}$$
$$= -1.3 - 2.5 - 1.3 - 1.6 = -6.7 \text{ kcal/mol}$$

The standard free energy for initiation is 1.8 kcal/mol so that the standard free energy change for formation of this DNA fragment is −4.9 kcal/mol. Similarly, $\Delta H° = -30.7$ kcal/mol and $\Delta S° = -85.8$ cal/(mol·K). The temperature dependence of the standard free energy change can be calculated by assuming the standard enthalpy change is independent of temperature. This relatively simple procedure permits the thermal stability of any linear DNA to be calculated to a good approximation.

Knowledge of the thermal properties of DNA fragments is important both physiologically and practically. Knowing the stability of DNA obviously is of interest in understanding genetic replication. From a practical standpoint, knowing the stability of DNA fragments is important in planning cloning experiments. DNA probes must be used that will form stable duplexes with the target DNA. The temperature at which the duplex becomes stable can be estimated by calculating the melting temperature, T_m, that is, the temperature at which half of the strands are in the double helix conformation. The equilibrium constant for formation of the double helix duplex is

$$K = [D]/[S]^2 \tag{3-21}$$

where D is the helical duplex and S is the single strand. If the total concentration of the oligonucleotide is C_0, the concentration of single strands at T_m is $C_0/2$ and that of the duplex is $C_0/4$. If we insert these relationships into Eq. 3-21, we see that $K = 1/C_0$ at T_m. If we insert the relationship between the equilibrium constant and the standard free energy change, we obtain

$$\Delta G°/RT_m = \ln C_0 \tag{3-22}$$

or

$$\Delta H°/RT_m - \Delta S°/R = \ln C_0$$

If Eq. 3-22 is solved for T_m, we obtain

$$T_m = \Delta H°/(\Delta S° + R \ln C_0) \tag{3-23}$$

For the case under discussion, if $C_0 = 0.1$ mM, $T_m = 295$ K.

In this brief discussion, we have neglected the fact that DNA is a polyelectrolyte due to the negatively charged phosphodiester backbone. The polyelectrolyte nature of DNA means that the ionic environment, particularly positively charged ions, strongly influences the structure and behavior of DNA. Normally a negatively charged polymer such as DNA would exist as a rod, because of the charge repulsion. However, we know that DNA is packaged into a small volume in cells. This involves the twisting of the double helix, the formation of loops, etc. Interactions with metal ions and positively charged proteins are necessary for this packaging to occur. Interested readers should consult more complete descriptions of nucleic acids for information on this interesting subject (3,7).

3.9 RNA

In principle, the structure of RNA can be discussed exactly as the discussion of DNA. The principles are the same, hydrogen bonding between bases, stacking interactions, and hydrophobic interactions determine the structure. Of course, a ribose is present, rather than deoxyribose, and uracil is substituted for thymine. In addition, several modified bases are commonly found in RNAs. Unfortunately, understanding and predicting RNA structures is more complex than for DNA. Ultimately, this is because there are several quite different biological functions for RNA and thus several quite different types of RNA. Generally, RNAs do not form intermolecular double helices. Instead, double helices are formed within an RNA molecule. These can be loops, hairpins, bulges, etc. Some idea of the diversity of structures that can be formed is shown in Figure 3-9, where an RNA molecule is shown, along with the different types of structures that can be formed.

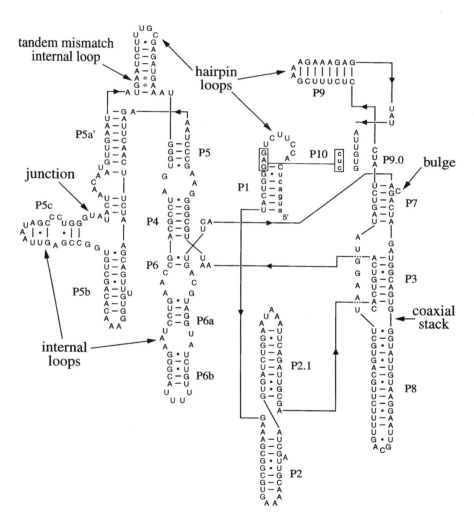

FIGURE 3-9. Proposed secondary structure of the group I intron of mouse derived *Pneumocystis carinii*. The areas indicate secondary structures within the intron, including base-paired helices. Reproduced from J. SantaLucia Jr. and D. H. Turner, *Biopolymers* **44**, 309 (1997). Copyright © 1997 Biopolymers. Reprinted by permission of John Wiley & Sons, Inc.

As with DNA, the RNA structures can be predicted reasonably well by considering nearest-neighbor interactions. However, because of the diversity of the structures, the models are more complex. Nevertheless, quite reasonable RNA structures can be predicted using the thermodynamics derived from simple systems. We will not delve further into RNA structures here, except to say that the principles of model building have been developed sufficiently here so that the interested reader can proceed directly to current literature on this subject (11–14). As with DNA, metal ions play an important role in the biological packaging of RNA.

Finally, the interactions between RNA and DNA are of obvious physiological relevance. These have been studied quite extensively, but they are not nearly as well understood as the interactions between DNA fragments and within RNA molecules. Simple models are not yet available to calculate the properties of DNA–RNA structures. At this point we will leave our discussion of thermodynamics of biological systems although many more interesting examples could be discussed.

REFERENCES

1. P. Mitchell, *Nature* **191**, 144 (1961).
2. T. G. Dewey and G. G. Hammes, *J. Biol. Chem.* **256**, 8941 (1981).
3. C. R. Cantor and P. R. Schimmel, *Biophysical Chemistry*, W. H. Freeman, New York, 1980.
4. L. Stryer, *Biochemistry,* 4th edition, W. H. Freeman, New York, 1995.
5. C. Branden and J. Tooze, *Introduction to Protein Structure,* 2nd edition, Garland Publishing, New York, 1999.
6. W. Kauzmann, *Adv. Protein Chem.* **14**, 1 (1959).
7. J. D. Watson, N. H. Hopkins, J. W. Roberts, J. A. Steitz, and A. M. Weiner, *Molecular Biology of the Gene,* 4th edition, Benjamin-Cummings, Menlo Park, CA, 1987.
8. K. J. Breslauer and J. M. Sturtevant, *Biophys. Chem.* **7**, 205 (1977).
9. J. SantaLucia, Jr., H. T. Allawi, and P. A. Seneviratne, *Biochemistry* **35**, 3555 (1996).
10. P. N. Borer, B. Dengler, I. Tinoco, Jr., and O. C. Uhlenbeck, *J. Mol. Biol.* **86**, 843 (1974).
11. I. Tinoco, Jr., O. C. Uhlenbeck, and M. D. Levine, *Nature* **230**, 362 (1971).
12. J. SantaLucia, Jr. and D. H. Turner, *Biopolymers* **44**, 309 (1997).
13. T. Xia, J. SantaLucia, Jr., M. E. Burkard, R. Kierzek, S. J. Schroeder, X. Jiao, C. Cox, and D. H. Turner, *Biochemistry* **37**, 14719 (1998).
14. M. Burkard, D. H. Turner, and I. Tinoco, Jr., *The RNA World,* 2nd edition, Cold Spring Harbor Laboratory Press, Cold Spring Harbor, NY, 1999, p. 233.

PROBLEMS

3-1. **A.** Glutamine is an important biomolecule made from glutamate. Calculate the equilibrium constant for the reaction

$$\text{Glutamate} + NH_3 \rightleftharpoons \text{Glutamine} + H_2O$$

Use the standard free energies of formation in Appendix 1 to obtain the standard free energy change for this reaction. Under physiological conditions, the concentration of ammonia is about 10 mM. Calculate the ratio [glutamine]/[glutamate] at equilibrium. (By convention, the concentration of water is set equal to 1 since its concentration does not change significantly during the course of the reaction.)

B. Physiologically, glutamine is synthesized by coupling the hydrolysis of ATP to the above reaction:

$$\text{Glutamate} + NH_3 + ATP \rightleftharpoons \text{Glutamine} + ADP + P_i$$

Calculate $\Delta G°$ and the equilibrium constant for this reaction under standard conditions (see Appendix 1). Assume that NH_3 and P_i are maintained at about 10 mM and that [ATP]/[ADP] = 1. What is the ratio of glutamate to glutamine at equilibrium? What ratio is needed to convert glutamate to glutamine spontaneously, that is, to make $\Delta G < 0$?

Do all calculations at 1 atmosphere and 298 K. Although this is not the physiological temperature, the results are not significantly altered.

3-2. Adipose tissues contain high levels of fructose. Fructose can enter the glycolytic pathway directly through the reaction

$$\text{Fructose} + ATP \rightleftharpoons \text{Fructose-6-phosphate} + ADP$$

Assume that the standard free energy change for this reaction with the same standard state used in Table 3-1 is -17.0 kJ/mol. If fructose is substituted for glucose in "glycolysis," what would the overall reaction be for the conversion of fructose to 2-glyceraldehyde, Part 1 of glycolysis? What would the overall reaction be for the complete metabolic cycle? Calculate $\Delta G°$ and ΔG for these two reactions. Assume the concentration of fructose is 5.0 mM, and the concentrations of the other metabolites are as in Table 3-2.

3-3. Glucose is actively transported into red blood cells by coupling the transport with the hydrolysis of ATP. The overall reaction can be written as

$$ATP + H_2O + n\,\text{Glucose(outside)} \rightleftharpoons n\,\text{Glucose(inside)} + ADP + P_i$$

If the ratio of [ATP]/[ADP] = 1 and $[P_i]$ = 10 mM, what is the concentration gradient, [glucose(inside)]/[glucose(outside)], that is established at 298 K? Calculate this ratio for $n = 1$, $n = 2$, and $n = n$. Does this suggest a method for determining the value of n? Use the value of $\Delta G°$ for the hydrolysis of ATP in Appendix 2 for these calculations.

3-4. Derive the equation for the temperature dependence of the standard free energy change for protein denaturation when ΔC_p is not equal to zero. As the starting temperature in the derivation, use the higher temperature at which $\Delta G° = 0$.

Hint: The easiest way to proceed is to calculate the temperature dependence of $\Delta H°$ and $\Delta S°$. These relationships can then be combined to give the temperature dependence of $\Delta G°$. The parameters in the final equation should be ΔC_P, the temperature, T, the temperature at which $\Delta G° = 0$, T_m, and the enthalpy change at the temperature where $\Delta G° = 0$, $\Delta H°_{T_m}$.

3-5. Specific genes in DNA are often searched for by combining a radioactive oligonucleotide, O, with the DNA that is complementary to a sequence in the gene being sought. This reaction can be represented as

$$O + DNA \rightleftharpoons Double\ strand$$

For such experiments, the concentration of the oligonucleotide is much greater than that of the specific DNA sequence. Assume that the probe is 5′-GGGAT-CAG-3′.

A. Calculate the equilibrium constant at 298 K for the interaction of the probe with the complementary DNA sequence using the parameters in Table 3-9.

B. Calculate the fraction of the DNA present as the double strand formed with the probe if the concentration of the probe is 1.0×10^{-4} M and the temperature is 298 K.

C. Find the melting temperature for this double strand if the concentration of the complementary DNA sequence is 1.0 nM. The melting temperature in this case is when [double strand]/[DNA] = 1.

3-6. For an electrochemical cell (e.g., a battery), the reversible work is $-nF\mathscr{E}$, where n is the number of moles of electrons involved in the chemical reaction, F is the Faraday, and $\mathscr{E}$ is the reversible voltage. This relationship is useful for considering coupled oxidation–reduction reactions in biochemical systems.

A. Use this relationship and Eq. 2-49 to derive an equation relating the free energy change for the reaction and the voltage of the cell. Your final equation should contain the standard free energy change for the reaction, the concentrations of the reactants, and the electrochemical voltage, as well as constants.

B. The voltage at equilibrium is usually designated as $\mathscr{E}°$. How is this related to the standard free energy change for the reaction? How might the equilibrium constant for a biochemical reaction be determined from voltage measurements? For the reaction

$$Malate + NAD^+ \rightleftharpoons Oxaloacetate + NADH + H^+$$

the voltage at equilibrium is –0.154 volts at 298 K. Calculate the equilibrium constant for this reaction. ($F = 96,485$ coulomb/mole and $n = 2$ for this reaction.)

CHAPTER 4

Chemical Kinetics

4.1   INTRODUCTION

Thermodynamics tells us what changes in state can occur, that is, the relative stability of states. For chemical reactions, it tells us what reactions can occur spontaneously. However, thermodynamics does not tell us the time scale for changes in state or how the changes in state occur; it is concerned only with the differences in the initial and final states. In terms of chemical reactions, it does not tell us *how* the reaction occurs, in other words, the molecular interactions that take place as a reaction occurs. For biological reactions, the rates are critical for the survival of the organism, and a primary interest of modern biology is the molecular events that lead to reaction. The study of the rates and mechanisms of chemical reactions is the domain of *chemical kinetics*. Thermodynamics provides no intrinsic information about mechanisms.

Many examples exist with regard to the importance of the time scale for chemical reactions. For those of you having diamond jewelry, you may be unhappy to know that the most stable state of carbon under standard conditions is graphite. So as you read this, your diamond is turning to graphite, but fortunately the time scale for this conversion is many hundreds of years. If graphite is more stable, why were diamonds formed? The answer is that the formation of diamonds did not occur under standard conditions: It is well known that at very high temperatures and pressures, graphite can be converted to diamond. In the biological realm, one of the most critical reactions is the hydrolysis of ATP to ADP and P_i. Thermodynamics tells us that the equilibrium lies far toward ADP and P_i. Yet solutions of ATP are quite stable under physiological conditions in the absence of the enzyme ATPase. A small amount of this enzyme will cause rapid and almost complete hydrolysis. Virtually all metabolic reactions occur much too slowly to sustain life in the absence of enzymes. Enzymes serve as catalysts and cause reactions to occur many orders of magnitude faster. Studies of the rates of hydrolysis under varying conditions allow us to say by what mechanism the reaction may occur, for example, how the substrates and products interact with the enzyme. Understanding the mechanism of biological processes is a research area of great current interest.

In this chapter, we will be interested in understanding some of the basic principles of chemical kinetics, with a few examples to illustrate the power of chemical kinetics. We will not delve deeply into the complex mathematics that is sometimes necessary, nor the many specialized methods that are sometimes used to analyze chemical kinetics. Many treatises are available that provide a more complete discussion of chemical kinetics (1–4). Because the examples used will be relatively simple, they will not necessarily involve biological processes. The principles discussed, however, are generally applicable to all systems. The background presented here should be sufficient to get you started on utilizing chemical kinetics and reading the literature with some comprehension. The key three concepts that will be discussed first are *rates of chemical reactions, elementary reactions,* and *mechanisms of chemical reactions.*

As a simple illustration, consider the reaction of hydrogen and iodine to give hydrogen iodide in the gas phase:

$$H_2 + I_2 \rightleftharpoons 2\,HI \qquad (4\text{-}1)$$

A possible mechanism for this reaction is for hydrogen and iodine to collide to produce hydrogen iodide directly. Another possible mechanism is for molecular iodine to first dissociate into iodine atoms and for the iodine atoms to react with hydrogen to produce hydrogen iodide. These possibilities can be depicted as

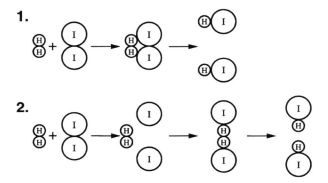

These two possible modes of reaction are quite distinct in terms of the chemistry occurring and are examples of two different mechanisms. We can also write these mechanisms in a more conventional manner:

1. $H_2 + I_2 \rightleftharpoons 2\,HI$ Elementary step and balanced chemical reaction   (4-2)

2. $I_2 \rightleftharpoons 2\,I$ Elementary step

   $2\,I + H_2 \rightleftharpoons 2\,HI$ Elementary step   (4-3)

   _____

   $H_2 + I_2 \rightleftharpoons 2\,HI$ Balanced chemical reaction

As indicated, the individual steps in the mechanism are called *elementary steps,* and they must always add up to give the balanced chemical reaction. In the first mechanism, the elementary reaction and balanced chemical reaction happen to be the same, but they are conceptually quite distinct, as we shall amplify later.

A very important point to remember is that a balanced chemical reaction gives no information about the reaction mechanism. This is not so obvious in the above example, but consider the reaction

$$3\ Fe^{2+} + HCrO_4^- + 7\ H^+ \rightleftharpoons 3\ Fe^{3+} + Cr^{3+} + 4\ H_2O$$

This clearly cannot be the reaction mechanism, as it would require the 11 reactants to encounter each other simultaneously, a very unlikely event.

4.2 REACTION RATES

The rate of a chemical reaction is a measure of how fast the concentration changes. If the concentration changes an amount Δc in time interval Δt, the rate of the reaction is $\Delta c/\Delta t$. If the limit of smaller and smaller time intervals is taken, this becomes dc/dt. If we apply this definition to Eq. 4-1, starting with H_2 and I_2 as reactants, the rate could be written as $-d[H_2]/dt$, $-d[I_2]/dt$, or $+d[HI]/dt$. For every mole of hydrogen and iodine consumed, 2 moles of hydrogen iodide are formed, so that this definition of the rate does not provide a unique definition: The value of the rate depends on which reaction component is under consideration. The rate of appearance of HI is twice as great as the rate of disappearance of H_2 and I_2. This ambiguity is not convenient, so a convention is used to define the reaction rate, namely, the rate of change of the concentration divided by its coefficient in the balanced chemical reaction. This results in a unique reaction rate, R, for a given chemical reaction. By convention, R is always positive. For the case under consideration,

$$R = -\frac{d[H_2]}{dt} = -\frac{d[I_2]}{dt} = \frac{1}{2}\frac{d[HI]}{dt}$$

Consider a more complex chemical equation:

$$2\ N_2O_5 \rightarrow 4\ NO_2 + O_2$$

In this case the reaction rate is

$$R = -\frac{1}{2}\frac{d[N_2O_5]}{dt} = \frac{1}{4}\frac{d[NO_2]}{dt} = \frac{d[O_2]}{dt}$$

Measuring the concentration as it changes with time can be very difficult, and some of our greatest advances in understanding chemical reactions have resulted from the development of new techniques for measuring the rates of chemical reactions, particularly the rates of very fast reactions. The most convenient method of measuring reaction rates is to mix the reactants together and to measure the subsequent change in concentrations continuously using a spectroscopic technique such as light absorption. It is sometimes not possible to find a physical property to monitor continuously, so it may be necessary to stop the reaction and measure the concentration chemically or through radioactive tracers. We will not dwell on this point, except to stress that measuring the rate of a chemical reaction may not be trivial.

Once a method has been established for determining the reaction rate, the next step is to measure the dependence of the reaction rate on the concentrations of the reactants. The dependence of the rate on the concentrations is called the *rate law*. *The rate law cannot be predicted from the balanced chemical equation. It must be determined experimentally.* A few examples will serve to illustrate this point. Consider the hydrolysis of ATP catalyzed by an ATPase. The overall reaction is

$$ATP \rightarrow ADP + P_i \tag{4-4}$$

and the observed dependence of the rate on concentration of ATP and enzyme in some cases is

$$R = \frac{k[E_0]}{1 + K_m/[ATP]} \tag{4-5}$$

where $[E_0]$ is the total enzyme concentration and k and K_m are constants. The rate law certainly cannot be deduced from the overall chemical reaction. In fact, the rate law is not the same for all ATPases. Many different ATPases are found in biological systems, and they do not all hydrolyze ATP by the same mechanism. As another example, consider the simple reaction

$$H_2 + Br_2 \rightarrow 2\,HBr \tag{4-6}$$

The experimentally determined rate law is

$$R = \frac{k[H_2][Br_2]^{1/2}}{1 + K[HBr]/[Br_2]} \tag{4-7}$$

As a final example, consider the redox reaction

$$5\,Br^- + BrO_3^- + 6\,H^+ \rightarrow 3\,Br_2 + 3\,H_2O \tag{4-8}$$

The observed rate law is

$$R = k[Br^-][BrO_3^-][H^+]^2 \tag{4-9}$$

These examples should emphasize the futility of attempting to predict the rate law from the balanced chemical equation.

The exponent of the concentration in the rate law is the *reaction order* with respect to that component (i.e., first order, second order, etc.). In some cases, such as Eq. 4-9, the concept of reaction order has a simple meaning. The rate law is first order with respect to Br^- and BrO_3^-, and second order with respect to H^+. For more complex rate laws such as Eqs. 4-5 and 4-7, the concept of reaction order cannot be used for all of the concentrations, except in limiting conditions. For Eq. 4-7, the rate law is 1/2 order with respect to Br_2 at high concentrations of Br_2 and/or low concentrations of HBr. The lowercase k's in the rate laws are called *rate constants*. The dimensions of rate constants can be deduced from the rate law by remembering that the rate is usually measured as M/s. In Eq. 4-9, k, therefore, has the dimensions of $M^{-3} s^{-1}$. The capital K's in the above equations also are constants, but they are combinations of rate constants rather than individual rate constants.

4.3 DETERMINATION OF RATE LAWS

Many different methods exist for determining rate laws. Only a few methods are considered here. With the routine use of computers, numerical integration of the differential equations and simultaneous fitting of the data are possible. However, as with any experimental approach, it is best to first fit the data to simple models before embarking on complex computer fitting. Computer programs will always fit the data, but it is important to be sure that the fit is a good one and the proposed rate law makes sense.

Probably the simplest and still often used method is the determination of initial rates. With this method, the rate of the reaction is measured at the very beginning of the reaction under conditions where the decrease in the concentrations of the reactants is so small that their concentrations in the rate law can be assumed to be the starting concentrations. In practice, this means that the concentrations should not change by more than a few percent. To illustrate the method, let's assume the rate law is

$$R = k(c_1)^a(c_2)^b \tag{4-10}$$

If c_2 is held constant and the initial rate is measured for different concentrations of c_1, the coefficient a can be determined. For example, if the concentration of c_1 is doubled and the rate increases by a factor of four, a must be equal to two. The same type of experiment can be done to determine b, namely, c_1 is held constant and c_2 is varied. The rate constant can readily be calculated, once the coefficients a and b are known, from the relationship $k = R/[(c_1)^a(c_2)^b]$. The determination of initial rates is especially useful for studying enzymatic reactions as we shall see later.

One of the most useful methods for determining the rate law is integration of the rate equation to obtain an analytical expression for the time dependence of the concentrations. The analytical equation is then compared with experimental data to see if it accurately describes the time dependence of the concentrations. As mentioned previously, computers can perform this integration numerically. We will consider two examples of integrated rate equations to illustrate the method.

The decomposition of nitrogen pentoxide can be written as

$$N_2O_5 \rightarrow 2\,NO_2 + \frac{1}{2}\,O_2 \tag{4-11}$$

Assume that the decomposition is a first order reaction:

$$R = -\frac{d[N_2O_5]}{dt} = k[N_2O_5] \tag{4-12}$$

This equation can easily be integrated

$$-\frac{d[N_2O_5]}{[N_2O_5]} = k\,dt$$

or

$$-\int_{[N_2O_5]_0}^{[N_2O_5]} \frac{d[N_2O_5]}{[N_2O_5]} = \int_0^t k\,dt$$

and

$$\ln([N_2O_5]/[N_2O_5]_0) = -kt \tag{4-13}$$

In Eq. 4-13, $[N_2O_5]_0$ is the concentration when $t = 0$. As illustrated in Figure 4-1, this equation predicts that a plot of $\ln[N_2O_5]$ versus t should be a straight line with a slope of $-k$. The data, in fact, conform to this rate equation under most conditions.

This analysis can also be carried out for higher order reactions. Hydrogen iodide will react to give hydrogen and iodine under certain conditions:

$$2\,HI \rightarrow H_2 + I_2 \tag{4-14}$$

The rate law is

$$R = -\frac{1}{2}\frac{d[HI]}{dt} = k[HI]^2 \tag{4-15}$$

Rearrangement of this equation gives

$$-\frac{d[HI]}{[HI]^2} = 2k\,dt$$

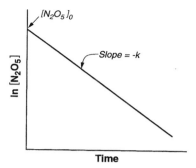

FIGURE 4-1. Demonstration of first order kinetics. Plot of $\ln[N_2O_5]$ versus the time according to Eq. 4-13. The straight line has a slope of $-k$.

which can be integrated to give

$$1/[HI] - 1/[HI]_0 = 2kt \tag{4-16}$$

Here $[HI]_0$ is the concentration when $t = 0$. As illustrated in Figure 4-2, this equation predicts that a plot of $1/[HI]$ versus t should be a straight line with a slope of $2k$. Again, the experimental data conform to this prediction.

It is frequently, but not always, possible to integrate rate equations analytically. Good experimental design can help to make the rate law relatively simple and therefore easy to integrate. In fact, experienced researchers will try to make the reaction first order whenever possible. This might seem like a major restriction but it is not. It is often possible to convert complex rate laws to conform to *pseudo first order kinetics*. For example, assume the rate law is

$$R = k(c_1)f(c_2)$$

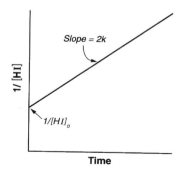

FIGURE 4-2. Demonstration of second order kinetics. Plot of $1/[HI]$ versus time according to Eq. 4-16. The straight line has a slope of $2k$.

where $f(c_2)$ is a function of the concentration of c_2. The function could be very complex or a simple power of the concentration—it does not matter. If the concentration of c_2 is made much larger than the concentration of c_1, then it can be assumed to remain constant throughout the reaction, and the rate law becomes

$$R = k'(c_1)$$

where $k' = kf(c_2)$. The "constant" $kf(c_2)$ is called a "pseudo" first order rate constant since it is constant under the experimental conditions used, but actually depends on the concentration c_2. If the concentration of c_2 is varied, the dependence of $f(c_2)$ on c_2 can be determined. It might seem restrictive to have a high concentration of c_2, but in practice it is often sufficient to have the concentration of c_2 only about a factor of 10 higher than c_1. Careful experimental design can make the job of determining the rate law much easier!

In determining the rate law, it is often necessary to use a broad range of experimental conditions before trying to interpret the rate law in terms of a chemical mechanism. As an illustration consider the reaction

$$I^- + OCl^- \rightarrow OI^- + Cl^- \tag{4-17}$$

At constant pH, the rate law determined experimentally is

$$R = k[I^-][OCl^-] \tag{4-18}$$

If the pH is varied, it is found that k also varies. It was determined that $k = k'/[OH^-]$, where k' is a constant. Thus, a more complete rate law is

$$R = k'[I^-][OCl^-]/[OH^-] \tag{4-19}$$

The more information that can be determined about the concentration dependence of the rate, the better the mechanism that can be postulated.

4.4   RADIOACTIVE DECAY

A good example of a first order rate process is radioactive decay. Radioactive isotopes are frequently used in biological research. For example, the radioactive isotope of phosphorus, ^{32}P, gives off radiation according to the reaction

$$^{32}P \rightarrow {}^{32}S + \beta^- \tag{4-20}$$

where β^- is a high-energy electron. The rate law for radioactive decay is

$$-\frac{d[^{32}P]}{dt} = k[^{32}P] \tag{4-21}$$

Integration of this rate law as done previously (Eq. 4-13) gives

$$[^{32}P] = [^{32}P]_0 \, e^{-kt} \qquad\qquad (4\text{-}22)$$

or

$$\ln([^{32}P]/[^{32}P]_0) = -kt$$

The rate of radioactive decay is usually given in terms of the half-life of the radioactive decay; in this case, the half-life is 14.3 days. In terms of the integrated rate law, Eq. 4-22, when half of the original radioactivity has decayed, $\ln(1/2) = -kt$, or the half-life for decay is $t_{1/2} = (\ln 2)/k$.

Thus, for radioactive decay, the half-life is constant. It does not matter when we start counting, or how much radioactivity we start with, the radioactivity will decay to half of its original value in 14.3 days. This is a special property of first order rate processes and is not true for other reaction orders where the half-life depends on the concentrations of the reactants. This is very convenient because regardless of when we start observing the rate of reaction, at $t = 0$, or at $t =$ any value, the integrated rate law is a simple single exponential.

4.5 REACTION MECHANISMS

In general, many mechanisms are possible for a given reaction. Mechanisms are proposals for how the reaction occurs. A proposed mechanism must be consistent with the experimentally observed rate law, but this is usually true for many mechanisms. *Kinetic studies can disprove a mechanism but cannot prove a mechanism.* As a practical matter the simplest mechanism consistent with all of the data is most appropriate, but at the end of the day, all that can be said is a specific mechanism is consistent with known data. It is not possible to say that this *must* be "the" mechanism.

A mechanism consists of a combination of *elementary steps*, which must sum up to give the overall reaction. For an elementary step, the order and molecularity, the number of molecules involved in the reaction, are the same. Therefore, for elementary steps, the rate law can be written as the product of the concentrations of all reactants, each raised to the power of their stoichiometric coefficient, multiplied by a rate constant. Some examples of elementary steps and associated rates laws are given below.

$$H_2 + I_2 \rightarrow 2\,HI \qquad R = k[H_2][I_2]$$

$$2\,I + H_2 \rightarrow 2\,HI \qquad R = k[I]^2[H_2]$$

$$O_3 \rightarrow O_2 + O \qquad R = k[O_3]$$

Remember rate laws can be derived from the chemical equation *only* for elementary steps, and never for the balanced chemical equation of the overall reaction. If an elementary reaction is reversible, then the rate law is the difference between the rates of the forward and reverse reactions. Therefore, for the elementary step

$$2\,A + B \underset{k_r}{\overset{k_f}{\rightleftharpoons}} C + D$$

$$R = k_f[A]^2[B] - k_r[C][D]$$

We now have two criteria for a possible mechanism: (1) It must be consistent with the observed rate law, and (2) the elementary steps must add up to give the overall balanced chemical reaction. Let us return to the two proposed mechanisms for the reaction of H_2 and I_2, Eqs. 4-2 and 4-3. The first mechanism contains only a single elementary step so that the rate law is

$$R = k_f[H_2][I_2] - k_r[HI]^2 \tag{4-23}$$

The second mechanism contains two elementary steps so that some assumptions must be made to derive the rate law. We will assume that iodine atoms are in rapid equilibrium with molecular iodine, or to be more specific that this equilibrium is adjusted much more rapidly than the reaction of iodine atoms with molecular hydrogen. Furthermore, the concentration of iodine atoms is assumed to be much less than that of molecular iodine. These assumptions are, in fact, known to be correct. If we now consider the second elementary step in the mechanism,

$$2\,I + H_2 \underset{k_2}{\overset{k_1}{\rightleftharpoons}} 2\,HI$$

the rate law can be written as

$$R = k_1[I]^2[H_2] - k_2[HI]^2 \tag{4-24}$$

but since the first step is always at equilibrium throughout the course of the reaction, $[I]^2 = K[I_2]$, where K is the equilibrium constant for the dissociation reaction. Substituting this relationship into Eq. 4-24 gives

$$R = k_1 K[I_2][H_2] - k_2[HI]^2 \tag{4-25}$$

Equations 4-23 and 4-25 are identical in form as only the definitions of the constants are different, $k_f = k_1 K$ and $k_r = k_2$. Thus, we have shown that both mechanisms are consistent with the rate law, and therefore both are possible mechanisms. Even for this simple reaction, there remains a debate as to which is the more likely mechanism.

It is easy to postulate a third possible mechanism with the following elementary steps:

$$H_2 \rightleftharpoons 2\,H$$

$$2\,H + I_2 \rightleftharpoons 2\,HI$$

By analogy with the mechanism involving iodine atoms, it can be seen that this mechanism would give the rate law of Eq. 4-25 with K now being the dissociation constant for molecular hydrogen in equilibrium with hydrogen atoms. However, K is a known constant and if this constant is combined with the value of k_1K determined experimentally, k_1 can be calculated. It is found that the value of k_1 is much larger than the rate constant characterizing the maximum rate at which two hydrogen atoms and molecular iodine encounter each other in the gas phase. Therefore, this mechanism can be ruled out as inconsistent with well established theory. Thus, we have a third criterion for disproving a mechanism.

As a second example of how mechanisms can be deduced, let us return to the reaction of I^- and OCl^- to produce OI^- and Cl^-, Eq. 4-17. The experimentally determined rate law is given by Eq. 4-19. What is a possible mechanism? One possibility is the following scheme:

$$OCl^- + H_2O \rightleftharpoons HOCl + OH^- \qquad \text{Fast, at equilibrium}$$

$$I^- + HOCl \xrightarrow{k_2} HOI + Cl^- \qquad \text{Slow} \qquad\qquad (4\text{-}26)$$

$$OH^- + HOI \rightleftharpoons H_2O + OI^- \qquad \text{Fast}$$

$$OCl^- + I^- \rightarrow OI^- + Cl^- \qquad \text{Overall reaction}$$

This mechanism is consistent with the balanced chemical equation. Now we must show that it is consistent with the observed rate law. The rate of a reaction is determined by the slowest step in the mechanism. Therefore, the rate of the overall reaction is given by the rate of the second elementary step:

$$R = k_2[I^-][HOCl] \qquad\qquad (4\text{-}27)$$

If the first step is assumed to be fast and the concentration of HOCl small relative that of I^- and Cl^-, then the concentration of HOCl can be calculated from the equilibrium relationship

$$K = [HOCl][OH^-]/[OCl^-]$$

$$[HOCl] = K[OCl^-]/[OH^-]$$

Insertion of this relationship into Eq. 4-27 gives

$$R = k_2K[I^-][OCl^-]/[OH^-] \qquad\qquad (4\text{-}28)$$

which is the observed rate law. Since this mechanism is consistent with the observed rate law and the balanced chemical equation, it is a possible mechanism.

You may very well be wondering if the creation of a mechanism is black magic. It is true that imagination and knowledge are important factors, but logic can be used. A species in the denominator results from a fast equilibrium prior to a rate determining step. Therefore, what needs to be found is a first step involving one of the reactants that produces the desired species, in this case OH^-. The other product of the first step must then react with the second reactant. The remainder of the steps are fast reactions that are necessary to produce a balanced chemical reaction. Note that none of the steps after the rate determining step play a role in determining the rate law. This is one of the simplest types of mechanisms, fast equilibria prior to and after a rate determining step. Nature is not always so obliging, and more complex mechanisms in which several steps occur at comparable rates are often necessary to account for experimental findings.

Here is another mechanism, quite similar in concept:

$$OCl^- + H_2O \rightleftharpoons HOCl + OH^- \qquad \text{Fast, at equilibrium}$$

$$I^- + HOCl \overset{k_2}{\rightarrow} ICl + OH^- \qquad \text{Slow} \qquad (4\text{-}29)$$

$$ICl + 2\,OH^- \rightleftharpoons OI^- + Cl^- + H_2O \quad \text{Fast}$$

$$\overline{}$$

$$OCl^- + I^- \rightarrow OI^- + Cl^- \qquad \text{Overall reaction}$$

Obviously this gives the same rate law as the mechanism in Eq. 4-26, as all events prior to the rate determining step are the same. However, the chemistry is quite different. In the first case iodide attacks the oxygen, in the second case it attacks the chlorine. These two mechanisms cannot be distinguished by kinetics. Both are equally consistent with the data.

Thus far we have considered mechanisms in which OCl^- is the initial reactant. Now let's look at mechanisms in which I^- is the initial reactant such as

$$I^- + H_2O \rightleftharpoons HI + OH^- \qquad \text{Fast, at equilibrium}$$

$$HI + OCl^- \rightarrow ICl + OH^- \qquad \text{Slow}$$

By analogy, it should be clear that this mechanism would give the correct rate law, and rapid reactions after the rate determining step can be added to give the correct balanced chemical equation. However, the equilibrium constant for the first step is known, and if this is combined with the results of the kinetic experiments, the rate constant for the rate determining step would exceed the theoretically possible value. This is because the concentration of HI is much, much smaller than that of HOCl. Thus, mechanisms of this type can be excluded on the basis of theoretical concepts.

These two examples illustrate how mechanisms can be related to the results of kinetic experiments. It is a great challenge to devise mechanisms. Once a mechanism is postulated, it is the job of the kineticist to devise experiments that will test the mechanism, often disproving it and requiring postulation of a new mechanism. However, it is important to remember that no matter how convincing the arguments, a mechanism cannot be proved. Thus far we have considered very simple reactions that are not biological, as biological reactions are typically very complex. The purpose of this discussion has been to illustrate the principles and concepts of chemical kinetics. We will later consider the kinetic analysis of enzymatic reactions, a very relevant and timely subject.

4.6   TEMPERATURE DEPENDENCE OF RATE CONSTANTS

Before discussing the kinetic analyses of biological reactions, a few additional concepts will be described. Reaction rates are often dependent on the temperature, and typically reactions go faster as the temperature increases. The first quantitative treatment of the temperature dependence of reaction rates was developed by Arrhenius in the late 1800s. He proposed that the temperature dependence of the rate constant could be described by the equation

$$k = A \exp(-E_a/RT) \tag{4-30}$$

where A is a constant, E_a is the *activation energy*, and T is the Kelvin temperature. This equation predicts that a plot of $\ln k$ versus $1/T$ should be a straight line with a slope of $-E_a/R$. This behavior is, in fact, followed in most cases. Equation 4-30 can be differentiated with respect to temperature to give

$$d(\ln k)/dT = E_a/RT^2 \tag{4-31}$$

This equation can be integrated to give

$$\ln(k_{T_2}/k_{T_1}) = \frac{(E_a/R)(T_2 - T_1)}{T_1 T_2} \tag{4-32}$$

Equation 4-32 permits the rate constant to be calculated at any temperature if it is known at one temperature and the activation energy has been determined. Note that these equations are similar in form to those describing the temperature dependence of the equilibrium constant except that the activation energy has replaced the standard enthalpy of reaction. In some cases, the activation energy may change with temperature, thereby making the analysis more complex.

The physical model behind the Arrhenius equation is that an energy barrier, E_a, must be overcome in order for the reaction to occur. This is shown conceptually in Figure 4-3. The reaction "path" or "coordinate" can be envisaged as the approach of the

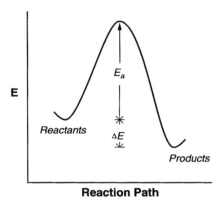

Reaction Path

FIGURE 4-3. Schematic representation of the energy, E, versus the "Reaction Path" during the course of a reaction as discussed in the text. The definition of the activation energy, E_a, is indicated.

reactants to each other that results in the lowest activation energy. The energy difference between products, P, and reactants, R, is ΔE. For simple reactions in the gas phase, the energy can be calculated as a function of the distance between reactants. These calculations define an energy surface, and the dynamic course of the reaction on this energy surface can be determined. Even in the gas phase, this can only be done for very simple reactions. For reactions in solution, this can only be considered a conceptual model.

You might guess from your knowledge of thermodynamics that the energy is probably not the best parameter to use to characterize the dynamics of a reaction. A theory has been developed that instead discusses the reaction path in terms of the free energy. This theory is called the *transition state theory*. The basic postulate is that a *transition state* is formed in the course of the reaction, which is in equilibrium with the reactants. This model is shown schematically in Figure 4-4. In the transition state theory, the reactants go through a transition state that is at a higher free energy than reactants. They must go over this free energy barrier to produce products. In Figure 4-4, the difference in free energy between products and reactants is $\Delta G°$, consistent with thermodynamic principles. The free energy difference between the transition state and reactants, $\Delta G°^{\ddagger}$, is called the standard free energy of activation. In terms of transition state theory, the rate constant can be written as

$$k = (k_B T/h) \exp(-\Delta G°^{\ddagger}/RT) \qquad (4\text{-}33)$$

where k_B is Boltzmann's constant and h is Planck's constant. (A more exact derivation includes an additional parameter, the transmission coefficient, which is usually equal to 1.) Since $\Delta G°^{\ddagger} = \Delta H°^{\ddagger} - T \Delta S°^{\ddagger}$, the rate constant can be restated as

$$k = (k_B T/h) \exp(\Delta S°^{\ddagger}/R) \exp(-\Delta H°^{\ddagger}/RT) \qquad (4\text{-}34)$$

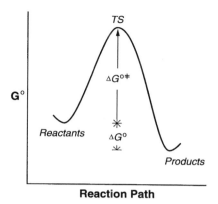

Reaction Path

FIGURE 4-4. Schematic representation of the standard free energy, G^0, versus the "Reaction Path" during the course of a reaction as discussed in the text. The free energy of activation, $\Delta G^{\circ\ddagger}$, is defined in this diagram. The transition state at the maximum of the free energy is indicated by TS. The reactants and products are at minima in the free energy curve.

Here $\Delta S^{\circ\ddagger}$ is the standard entropy of activation, and $\Delta H^{\circ\ddagger}$ is the standard enthalpy of activation.

If the standard entropy and enthalpy of activation are assumed to be temperature independent, Eq. 4-34 can be differentiated to give

$$d(\ln k)/dT = (\Delta H^{\circ\ddagger} + RT)/(RT^2) \tag{4-35}$$

Thus, the transition state theory predicts a temperature dependence of the rate constant very similar to the Arrhenius theory with $E_a = \Delta H^{\circ\ddagger} + RT$. Since RT is usually small relative to the standard enthalpy of activation, the activation energy and standard enthalpy of activation are usually quite similar. The Arrhenius and transition state formulations cannot be differentiated by this small difference in the temperature dependence since both the activation energy and the standard enthalpy of activation can be temperature dependent. From our knowledge of thermodynamics, we know that the enthalpy of activation is temperature dependent if a heat capacity difference exists between the transition state and the reactants. If the temperature dependence of the rate constant is to be analyzed in terms of transition state theory, it is more convenient to plot $\ln(k/T)$ versus $1/T$ as $d \ln(k/T)/dT = \Delta H^{\circ\ddagger}/(RT^2)$, or $d \ln(k/T)/d(1/T) = -\Delta H^{\circ\ddagger}/R$.

A simple derivation of Eq. 4-33 is possible. The concentration of the transition state, TS, can be calculated from the relationship

$$[TS]/[Reactants] = \exp(-\Delta G^{\circ\ddagger}/RT)$$

where [Reactants] represents the concentrations of the reactants raised to the appropriate powers for the stoichiometric equation of the elementary step. The rate of the

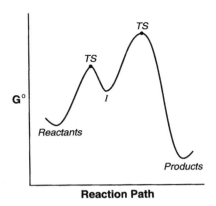

FIGURE 4-5. Schematic representation of the standard free energy, G^0, versus the "Reaction Path" during the course of a two-step reaction. The intermediate, I, is at a minimum in the free energy whereas the transition state, TS, for each step is at a maximum. The reactants and products are also at minima in the free energy curve.

reaction is the concentration of the transition state multiplied by the frequency with which it passes over the free energy barrier. This frequency of passage over the barrier can be derived from an analysis of the reaction coordinate with statistical mechanics and is given by $k_B T/h$. Therefore, the rate of reaction is

$$R = (k_B T/h)[\text{TS}] = [\text{Reactants}](k_B T/h)\exp(-\Delta G^{\circ\ddagger}/RT)$$

The rate constant in Eq. 4-33 follows directly from the above equation. This nonrigorous derivation provides a conceptual framework for the theory.

If a reaction goes through an intermediate, the intermediate would correspond to a minimum in the free energy, and each elementary step would have its own transition state. This is shown schematically in Figure 4-5 for a sequence of two elementary steps. The step with the highest free energy barrier is the rate determining step.

Transition state theory is a very useful method for correlating and understanding kinetic studies. Because the framework of the theory is similar to thermodynamics, this produces a consistent way of discussing chemical reactions. The entropy and enthalpy of activation are often discussed in molecular terms. It should be remembered that, for kinetics as with thermodynamics, such interpretations should be approached with extreme caution.

4.7 RELATIONSHIP BETWEEN THERMODYNAMICS AND KINETICS

Obviously the principles of thermodynamics and kinetics must be self-consistent. In fact, this places some useful restrictions on the relationships between rate constants. In order to see how rate constants are related to equilibrium constants, consider the elementary step

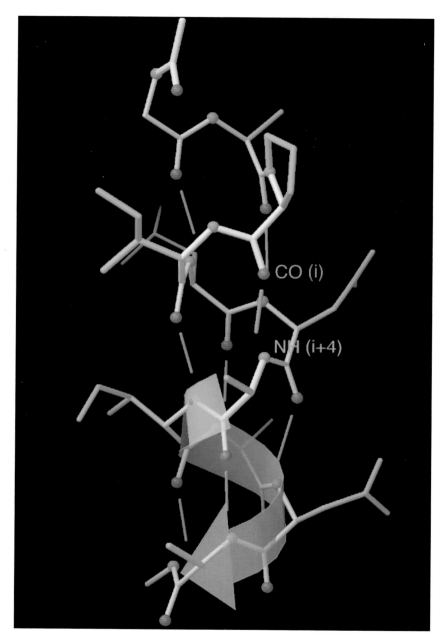

COLOR PLATE I. The α-helix found in many proteins. The yellow arrow follows the right-handed spiral of one helical turn. The hydrogen bonds between backbone peptide bonds are brown lines, the oxygens are red, and the nitrogens blue. The hydrogen bonds are formed between the *i*th carbonyl and the *i*+4 NH in the peptide backbone. Copyright by Professor Jane Richardson. Reprinted with permission.

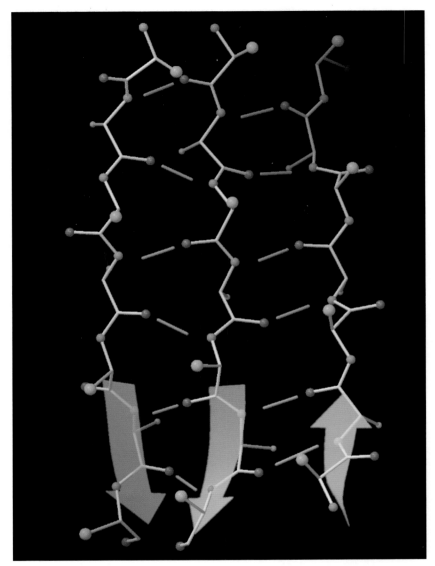

COLOR PLATE II. The ß-pleated sheets found in many proteins. Both parallel and antiparallel strand-strand interactions are shown, as indicated by the yellow arrows. Again, the hydrogen bonds between backbone peptide bonds are shown as brown lines between the oxygens and nitrogens. Copyright by Professor Jane Richardson. Reprinted with permission.

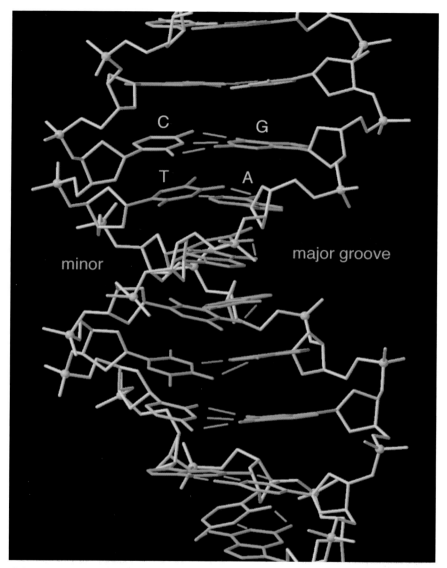

COLOR PLATE III. Stick-figure representation of the B form of the DNA double helix. The planes of the hydrogen-bonded bases can be seen, as well as the twisting of the chains to form a double helix. The pale yellow spheres are the phosphorous atoms of the sugar phosphate backbone on the outside of the helix. The bases are color coded, and the two grooves in the structure are labeled. Copyright by Professor Jane Richardson. Reprinted with permission.

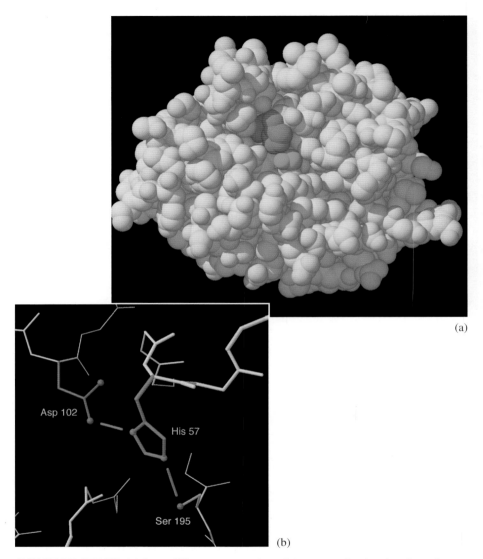

(a)

(b)

COLOR PLATE IV. (a) Space filling representation of chymotrypsin, showing the active site pocket where substrate binding occurs. The blue at the active site is imidazole, the green is serine, and the red is aspartate. (b) Arrangement of the protein residues at the active site of chymotrypsin. The serine is the residue acylated; the imidazole of the histidine serves as a general base catalyst; and the aspartate carboxyl group is hydrogen bonded to the imidazole. Copyright by Professor Jane Richardson. Reprinted with permission.

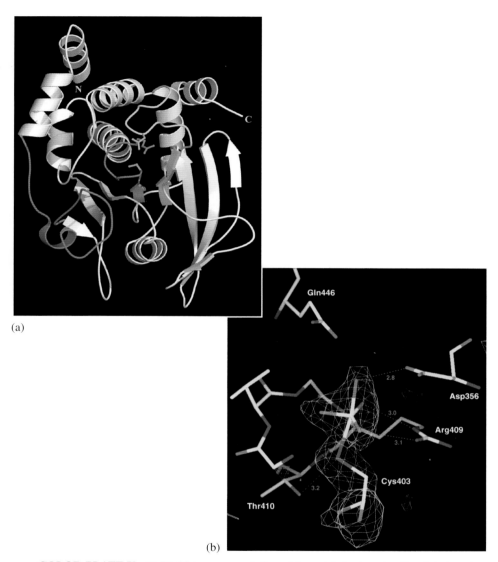

(a)

Gln446

2.8

Asp356

3.0

Arg409

3.1

Cys403

3.2

Thr410

(b)

COLOR PLATE V. (a) Backbone representation of the catalytic domain of protein tyrosine phosphatase. Coils and arrows represent α-helices and ß-strands, respectively. The cysteine (green), arginine (red), and aspartate (gold) are also shown. Reproduced from J. M. Denu, J. A. Stuckey, M. A. Saper, and J. E. Dixon, *Cell* **87**, 316 (1996). Copyright ©1996 Cell Press. reprinted with permission of Cell Press. (b) Vanadate bound at the active-site cysteine of protein tyrosine phosphatase. Other amino acids at the catalytic site are indicated, including the threonine and aspartate that participate in the catalytic mechanism. Reproduced from J. M. Denu, D. L. Lohse, J. Vijayalakshmi, M. A. Saper, and J. E. Dixon, *Proc. Natl. Acad. Sci. USA* **93,** 2493 (1996). Reprinted with permission of the Proceedings of the National Academy of Sciences USA. Reproduced by permission of the publisher via Copyright Clearance Center, Inc.

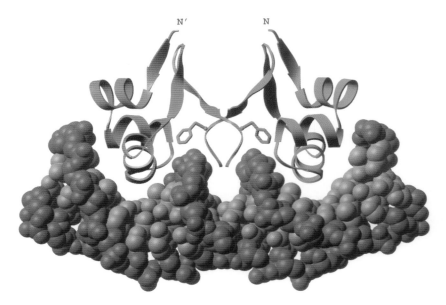

COLOR PLATE VI. Structure of the overall complex between Cro protein and λ operator DNA. The direction of the view is parallel with the major grooves of DNA and parallel with the recognition helices. Reproduced from R. A. Albright and B. W. Matthews, *J. Mol. Biol.* **280,** 137 (1998). Reprinted with permission from Academic Press, Inc.

COLOR PLATE VII. Schematic representations of the R (bottom, pink) and T (top, blue) forms of the hemoglobin $\alpha_2\beta_2$ tetramer. The hemes where oxygen binds can be seen in the structure. The yellow side chains form salt bonds in the T structure that are broken in the R structure. One pair of α-ß subunits also rotates with respect to the other by about 15° in the interconversation of R and T forms. Copyright by Professor Jane Richardson. Reprinted with permission.

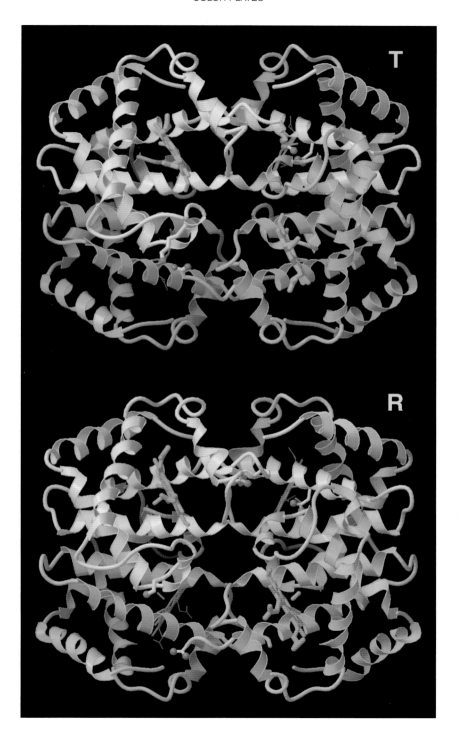

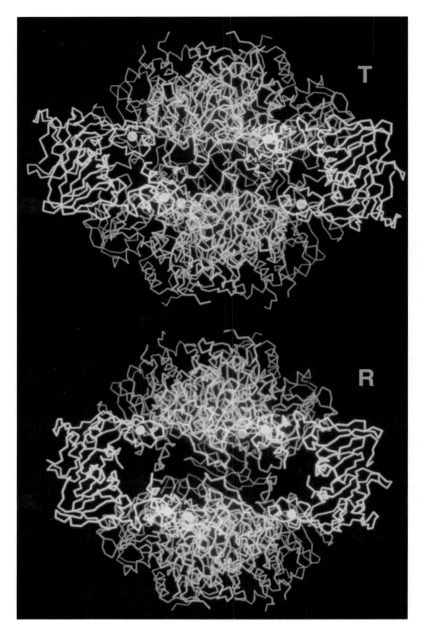

COLOR PLATE VIII. Cα backbone structures of the R (lower) and T (upper) states of aspartate transcarbamoylase. The two catalytic trimers are in green and are closer together in the T state. The three regulatory dimers (equatorial) are in yellow. One of the regulatory dimers is behind the large central cavity in this view. Adapted from W. N. Lipscomb, *Adv. Enzymol.* **68,** 67 (1994). Copyright © 1994 John Wiley & Sons, Inc. Reprinted by permission of John Wiley & Sons, Inc.

$$A + B \underset{k_2}{\overset{k_1}{\rightleftharpoons}} AB \tag{4-36}$$

The rate of this reaction is

$$R = -\frac{d[A]}{dt} = k_1[A][B] - k_2[AB] \tag{4-37}$$

At equilibrium the net rate of reaction must be zero. If $R = 0$, $k_1[A]_e[B]_e = k_2[AB]_e$, where the subscript e designates the equilibrium concentration. Thus, we see that

$$K = k_1/k_2 = [AB]_e/([A]_e[B]_e) \tag{4-38}$$

In this case, the equilibrium constant, K, is equal to the ratio of rate constants. Similar relationships between the rate constants and equilibrium constants can be found for more complex situations by setting the net rate equal to zero at equilibrium. If Eq. 4-38 is cast into the framework of transition state theory, we obtain

$$K = \exp[-(\Delta G_1^{\circ\ddagger} - \Delta G_2^{\circ\ddagger})/RT] = \exp(-\Delta G^\circ/RT) \tag{4-39}$$

This result indicates the relationship between the standard free energy changes of activation and the standard free energy change for the reaction. This relationship can also be seen in the diagram of the free energy versus reaction path.

A more subtle relationship can be found if reaction cycles occur because of the *principle of detailed balance* or *microscopic reversibility*. This principle states that a mechanism for the reaction in the forward direction must also be a mechanism for the reaction in the reverse direction. Furthermore, at equilibrium, the forward and reverse rates are equal along each reaction pathway. This means that once we have found a possible mechanism for the reaction in one direction, we have also found a possible mechanism for the reaction in the other direction.

To illustrate this principle, consider the following triangular reaction mechanism:

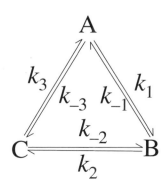

According to the principle of detailed balance, each of the individual reactions must be at equilibrium when equilibrium is attained, or

$$k_1[A]_e = k_{-1}[B]_e$$

$$k_2[B]_e = k_{-2}[C]_e$$

$$k_3[C]_e = k_{-3}[A]_e$$

If the right-hand sides of these equations are multiplied together and set equal to the left-hand sides of the equations multiplied together, we obtain

$$k_1 \, k_2 \, k_3 = k_{-1} \, k_{-2} \, k_{-3} \qquad (4\text{-}40)$$

Thus, we find that the six rate constants are not independent, nor are the three equilibrium constants! This result may seem obvious, but many people have violated the principle of detailed balance in the literature. It is important to confirm that this principle is obeyed whenever reaction cycles are present.

4.8   REACTION RATES NEAR EQUILIBRIUM

Before we consider the application of chemical kinetics to biological systems, we will discuss the special case of reaction rates near equilibrium. As we shall see, the rate laws become quite simple near equilibrium, and methods exist that permit very fast reactions to be studied near equilibrium. We will begin with an elementary step that is a reversible first order reaction, such as protein denaturation in the middle of the transition from the native to the denatured state:

$$N \underset{k_r}{\overset{k_f}{\rightleftharpoons}} D \qquad (4\text{-}41)$$

The rate equation characterizing this system is

$$-\frac{d[N]}{dt} = k_f[N] - k_r[D] \qquad (4\text{-}42)$$

We will now introduce new concentration variables; that is, we will set each concentration equal to its equilibrium value plus the deviation from equilibrium:

$$[N] = [N]_e + \Delta[N]$$

$$[D] = [D]_e + \Delta[D]$$

Note that the equilibrium concentrations are constants, independent of time, and mass conservation requires that $\Delta[N] = -\Delta[D]$. Inserting these relationships into Eq. 4-42 gives

$$-\frac{d[\Delta N]}{dt} = k_f([N]_e + \Delta[N]) - k_r([D]_e - \Delta[N])$$

$$= k_f[N]_e - k_r[D]_e + (k_f + k_r)\Delta[N]$$

$$= (k_f + k_r)\Delta[N] = \Delta[N]/\tau$$

In deriving the above relationship, use has been made of the relationship $k_f[N]_e = k_r[D]_e$ and $1/\tau = k_f + k_r$; τ is called the relaxation time. This first order differential equation can be integrated as before to give

$$\Delta[N] = \Delta[N]_0 e^{-t/\tau} \tag{4-43}$$

where $\Delta[N]_0$ is the deviation of N from its equilibrium value at $t = 0$.

Special note should be made of two points. First, the relaxation time can readily be obtained from experimental data simply by plotting the $\ln(\Delta[N])$ versus t, and second the first order rate constant characterizing this reaction, that is, the reciprocal relaxation time, is the sum of the two rate constants. If the equilibrium constant is known, both rate constants can be obtained from a single experiment.

How might such an experiment be carried out? In the case of protein denaturation, a small amount of denaturant such as urea might be added quickly to the solution. The ratio of the native and denatured protein would then move to a new equilibrium value characteristic of the higher urea concentration. Alternatively, if thermal denaturation is being studied, the temperature might rapidly be raised, establishing a new equilibrium ratio. The rate of conversion of the native state to the denatured state can be measured after the experimental conditions are changed. For a reversible first order reaction such as Eq. 4-41, the time course of the entire reaction follows a single exponential and the effective rate constant is the sum of the two rate constants. A wide range of methods exists for changing the equilibrium conditions: Concentration jumps and temperature jumps are particularly useful, but pressure jumps and electrical field jumps also have been used.

Studying rates near equilibrium is especially advantageous for higher order reactions and complex mechanisms. As a final example, we will consider the elementary step of an enzyme, E, combining with a substrate, S.

$$E + S \rightleftharpoons ES \tag{4-44}$$

The rate equation characterizing this system is

$$-\frac{d[\text{E}]}{dt} = k_\text{f}[\text{E}][\text{S}] - k_\text{r}[ES] \tag{4-45}$$

Again we will write the concentrations as the sum of the equilibrium concentration plus the deviation from equilibrium.

$$[\text{E}] = [\text{E}]_\text{e} + \Delta[\text{E}]$$

$$[\text{S}] = [\text{S}]_\text{e} + \Delta[\text{S}]$$

$$[ES] = [ES]_\text{e} + \Delta[ES]$$

Furthermore, mass conservation requires that $\Delta[\text{E}] = \Delta[\text{S}] = -\Delta[ES]$. Inserting these relationships into Eq. 4-45 gives

$$-\frac{d(\Delta[\text{E}])}{dt} = \{k_\text{f}([\text{E}]_\text{e} + [\text{S}]_\text{e}) + k_\text{r}\}\,\Delta[\text{E}] + k_\text{f}[\text{E}]_\text{e}[\text{S}]_\text{e} - k_\text{r}[ES]_\text{e} + k_\text{f}(\Delta[\text{E}])^2$$

Near equilibrium, the deviation of concentrations from their equilibrium values is so small that the term $(\Delta[\text{E}])^2$ can be neglected—this simplification in the rate equation results from being near equilibrium. For example, if the deviation is 10% (0.1), the square of the deviation is 1% (0.01). Furthermore, $k_\text{f}[\text{E}]_\text{e}[\text{S}]_\text{e} = k_\text{r}[ES]_\text{e}$, so that the final rate equation is

$$-\frac{d[\Delta\text{E}]}{dt} = \frac{\Delta[\text{E}]}{\tau} \tag{4-46}$$

with

$$1/\tau = k_\text{f}([\text{E}]_\text{e} + [\text{S}]_\text{e}) + k_\text{r} \tag{4-47}$$

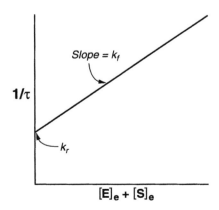

FIGURE 4-6. Schematic representation of a plot of the reciprocal relaxation time, $1/\tau$, versus the sum of the equilibrium concentrations, $[\text{E}]_\text{e} + [\text{S}]_\text{e}$, according to Eq. 4-47. As indicated, both of the rate constants can be obtained from the data.

Although the reaction of enzyme with substrate is a second order reaction, the rate equation near equilibrium, Eq. 4-46, is the same as for a first order reaction. In fact, all rate equations become first order near equilibrium! This is because only terms containing Δc are retained in the rate law: $(\Delta c)^2$ and higher powers are neglected. As before, Eq. 4-46 can easily be integrated, and the relaxation time can be obtained from the experimental data. If the relaxation time is determined at various equilibrium concentrations, a plot of $1/\tau$ versus $([E]_e + [S]_e)$ can be made, as shown schematically in Figure 4-6. The intercept is equal to k_r and the slope is equal to k_f. The study of reactions near equilibrium has been especially important for biological systems, as it has permitted the study of important elementary steps such as hydrogen bonding and protolytic reactions, as well as more complex processes such as enzyme catalysis and ligand binding to macromolecules.

REFERENCES

1. I. Tinoco, Jr., K. Sauer, and J. C. Wang, *Physical Chemistry: Principles and Applications in Biological Sciences,* 3rd edition, Prentice Hall, Englewood Cliffs, NJ, 1995.

2. C. R. Cantor and P. R. Schimmel, *Biophysical Chemistry*, W. H. Freeman, San Francisco, 1980.

3. G. G. Hammes, *Enzyme Catalysis and Regulation*, Academic Press, New York, 1982.

4. A. Fersht, *Structure and Mechanism in Protein Science: A Guide to Enzyme Catalysis and Protein Folding*, W. H. Freeman, San Francisco, 1999.

PROBLEMS

4-1. The activity of the antibiotic penicillin slowly decomposes when stored in a buffer at pH 7.0, 298 K. The time dependence of the penicillin antibiotic activity is given in the table below.

Time (weeks)	Penicillin Activity (arbitrary units)
0	10,100
1.00	8,180
2.00	6,900
3.00	5,380
5.00	3,870
8.00	2,000
10.00	1,330
12.00	898
15.00	403
20.00	167

What is the rate law for this reaction, that is, what is the order of the reaction with respect to the penicillin concentration? Calculate the rate constant from the data if possible. (Data adapted from Ref. 1.)

4-2. The kinetics of the reaction

$$2Fe^{3+} + Sn^{2+} \rightarrow 2Fe^{2+} + Sn^{4+}$$

has been studied extensively in acidic aqueous solutions. When Fe^{2+} is added initially at relatively high concentrations, the rate law is

$$R = k[Fe^{3+}]^2[Sn^{2+}]/[Fe^{2+}]$$

Postulate a mechanism that is consistent with this rate law. Show that it is consistent by deriving the rate law from the proposed mechanism.

4-3. The conversion of L-malate to fumarate is catalyzed by the enzyme fumarase:

The nonenzymatic conversion is very slow in neutral and alkaline media and has the rate law

$$R = k[\text{L-malate}]/[H^+]$$

Postulate two mechanisms for the nonenzymatic conversion and show that they are consistent with the rate law.

4-4. The radioactive decay rates of naturally occurring radioactive elements can be used to determine the age of very old materials. For example, $^{14}_{6}C$ is radioactive and emits a low-energy electron with a half-life of about 5730 years. Through a balance of natural processes, the ratio of $^{14}C/^{12}C$ is constant in living organisms. However, in dead organisms or material, this ratio decreases as the ^{14}C decays. Since the radioactive decay is known to be a first order reaction, the age of the material can be estimated by measuring the decrease in the $^{14}C/^{12}C$ ratio. Suppose a piece of ancient wool is found in which the ratio has been found to decrease by 20%. What is the age of the wool?

4-5. The nonenzymatic hydration of CO_2 can be written as

$$CO_2 + H_2O \rightleftharpoons H_2CO_3$$

The reaction is found to be first order in both directions. Because the water concentration is constant, it does not appear in the expression for the equilibrium or rate equation. The first order rate constant in the forward direction has a value of 0.0375 s^{-1} at 298 K and 0.0021 s^{-1} at 273 K. The thermodynamic parameters for the equilibrium constant at 298 K are $\Delta H° = 1.13$ kcal/mol and $\Delta S°$ $= -8.00$ cal/(mol·K).

A. Calculate the Arrhenius activation energy for the rate constant of the forward reaction. Also calculate the enthalpy and entropy of activation according to transition state theory at 298 K.

B. Calculate the rate constant for the reverse reaction at 273 and 298 K. Assume that $\Delta H°$ is independent of the temperature over this temperature range.

C. Calculate the Arrhenius activation energy for the rate constant of the reverse reaction. Again, also calculate the enthalpy and entropy of activation at 298 K.

4-6. A hydrogen bonded dimer is formed between 2-pyridone according to the reaction

The relaxation time for this reaction, which occurs in nanoseconds, has been determined in chloroform at 298 K at various concentrations of 2-pyridone. The data obtained are [G. G. Hammes and A. C. Park, *J. Am. Chem. Soc.* **91**, 956 (1969)]:

2-Pyridone (M)	$10^9 \tau$ (s)
0.500	2.3
0.352	2.7
0.251	3.3
0.151	4.0
0.101	5.3

From these data calculate the equilibrium and rate constants characterizing this reaction. *Hint*: If the expression for the relaxation time is squared, the concentration dependence can be expressed as a simple function of the total concentration of 2-pyridone.

CHAPTER 5

Applications of Kinetics to Biological Systems

5.1 INTRODUCTION

We will now consider some applications of kinetic studies to biological systems. This discussion will center on enzymes, as kinetic analyses of enzymes represent a major research field and have provided considerable insight into how enzymes work. Enzymes are proteins that are incredibly efficient catalysts: They typically increase the rate of a chemical reaction by six orders of magnitude or more. Understanding how this catalytic efficiency is achieved and the exquisite specificity of enzymes has intrigued biologists for many decades and still provides a challenge for modern research. Since enzyme deficiencies are the source of many diseases, enzymes are also a target for medical research and modern therapeutics. The field of enzyme kinetics and mechanisms is so vast that we will only present an abbreviated discussion. More complete discussions are available (cf. Refs. 1–3).

We will also discuss catalysis by RNA (ribozymes), which is important in the processing of RNA in biological systems, as well as some kinetic studies of DNA denaturation and renaturation.

5.2 ENZYME CATALYSIS: THE MICHAELIS–MENTEN MECHANISM

We now consider the analysis of a simple enzymatic reaction. This discussion will introduce some new concepts that are particularly useful for analyzing complex systems. The conversion of a single substrate to product will be taken as a prototype reaction:

$$S \rightarrow P \tag{5-1}$$

A typical example is the hydration of fumarate to give L-malate (and the reverse dehydration reaction) catalyzed by the enzyme fumarase. If a very low concentration of enzyme is used relative to the concentration of substrate, a plot of the initial rate of the reaction, or initial velocity, v, is hyperbolic, as shown in Figure 5-1. The limiting initial velocity at high concentrations of substrate is called the *maximum velocity*, V_m, and the concentration of substrate at which the initial velocity is equal to $V_m/2$ is called the *Michaelis constant*, K_M.

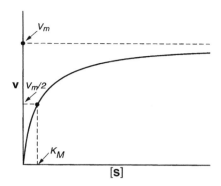

FIGURE 5-1. A plot of the initial velocity, v, versus the substrate concentration, [S], for an enzymatic reaction that can be described by the Michaelis–Menten mechanism.

A mechanism that quantitatively accounts for the dependence of the initial velocity on substrate concentration was postulated by Michaelis and Menten. It can be written in terms of elementary steps as

$$E + S \underset{k_2}{\overset{k_1}{\rightleftharpoons}} ES \xrightarrow{k_3} E + P \tag{5-2}$$

where E is the enzyme and ES is a complex consisting of the enzyme and substrate. The total concentration of enzyme, $[E_0]$, is assumed to be much less than the total concentration of substrate, $[S_0]$. This mechanism is an example of *catalysis* in that the enzyme is not consumed in the overall reaction, which is greatly accelerated by the enzyme. If we tried to do an exact mathematical analysis of the mechanism in Eq. 5-2, coupled differential equations would need to be solved. Fortunately, this is not necessary. We will consider two approximations that can adequately account for the data.

The first is the *equilibrium approximation*. With this approximation, the first step is assumed to be always at equilibrium during the course of the reaction. This means that the equilibration of the first step is much more rapid than the breakdown of ES to P, or $k_3 \ll k_2$. We then have

$$\frac{d[P]}{dt} = v = k_3[ES] \tag{5-3}$$

and

$$k_1/k_2 = [ES]/([E][S]) \tag{5-4}$$

Furthermore, conservation of mass requires that

$$[E_0] = [E] + [ES] \tag{5-5}$$

$$= [ES](1 + [E]/[ES]) = [ES]\{1 + k_2/(k_1[S])\}$$

or

$$[ES] = [E_0]/\{1 + k_2/(k_1[S])\} \tag{5-6}$$

Substitution of Eq. 5-6 into 5-3 gives

$$\frac{d[P]}{dt} = v = \frac{k_3[E_0]}{1 + k_2/(k_1[S])} \tag{5-7}$$

which can be rewritten as

$$v = \frac{V_m}{1 + K_M/[S]} \tag{5-8}$$

with

$$V_m = k_3[E_0] \tag{5-9}$$

and

$$K_M = k_2/k_1 \tag{5-10}$$

Equation 5-8 is found to account quantitatively for the data. The maximum velocity is proportional to the total enzyme concentration, and when $K_M = [S]$, $v = V_m/2$, as required.

In some cases, it is possible to obtain independent estimates or measurements of the equilibrium constant for the first step. Sometimes this independent measurement is in good agreement with the constant obtained from kinetic studies, sometimes it is not. Obviously, when the Michaelis constant and the equilibrium constant are not in agreement, the mechanism must be reexamined.

A more general analysis of the Michaelis–Menten mechanism makes use of the *steady-state approximation*. This approximation does not make any assumptions about the relative values of the rate constants but assumes that the concentrations of E and ES are very small relative to S, consistent with the experimental conditions. Under these conditions, it is assumed that the rate of change of the concentrations of E and ES is very small relative to the rate of change of the concentration of S, or

$$d[E]/dt = d[ES]/dt \approx 0$$

This is the steady-state approximation, an approximation that is very important for analyzing complex mechanisms. A careful mathematical analysis shows that this is a very good approximation under the experimental conditions used, namely, when the total substrate concentration is much greater than the total enzyme concentration.

What does the steady-state approximation mean? If we go back to the mechanism (Eq. 5-2), we find that

$$-\frac{d[ES]}{dt} = (k_2 + k_3)[ES] - k_1[E][S] \approx 0$$

or

$$k_1/(k_2 + k_3) = [ES]/([E][S]) \tag{5-11}$$

Note the similarity of this equation to the equilibrium constant (Eq. 5-4). This means that the ratio of concentrations remains constant. However, the ratio is not the equilibrium concentrations; it is the steady-state concentrations. If $k_3 \ll k_2$, the steady state is the same as the equilibrium state, so that the equilibrium condition is a special case of the steady state. For all other steady-state situations, the ratio of concentrations is always less than the equilibrium ratio.

We can now calculate the rate law for the steady-state approximation exactly as for the equilibrium approximation.

$$[E_0] = [ES](1 + [E]/[ES]) = [ES]\{1 + (k_2 + k_3)/(k_1[S])\} \tag{5-12}$$

$$v = \frac{d[P]}{dt} = k_3[ES] = \frac{k_3[E_0]}{1 + (k_2 + k_3)/(k_1[S])}$$

$$v = \frac{V_m}{1 + K_M/[S]} \tag{5-13}$$

with

$$V_m = k_3[E_0] \tag{5-14}$$

and

$$K_M = (k_2 + k_3)/k_1 \tag{5-15}$$

Thus, it is clear that the equilibrium approximation and steady-state approximation give rate laws that are indistinguishable experimentally. Most generally, the Michaelis constant is a steady-state constant, but in a limiting case it can be an equilibrium constant. As previously stated, both situations have been observed. The steady-state approximation is more general than the equilibrium approximation and is typically employed in the analysis of enzyme mechanisms.

With modern computers, data analysis is very easy, and experimental data can be fit directly to Eq. 5-8 (or Eq. 5-13) by a nonlinear least squares fitting procedure. However, it is always a good idea to be sure that the data indeed do conform to the equation of choice by a preliminary analysis. Equation 5-8 can be linearized by taking its reciprocal:

$$1/v = 1/V_m + (K_M/V_m)/[S] \tag{5-16}$$

This is called the Lineweaver–Burke equation. A plot of $1/v$ versus $1/[S]$ is linear, and the slope and intercept can be used to calculate the maximum velocity and Michaelis constant. This equation can be used for the final data analysis providing proper statistical weighting is used. An alternative linearization of Eq. 5-8 is to multiply Eq. 5-16 by [S] to give

$$[S]/v = [S]/V_m + K_M/V_m \tag{5-17}$$

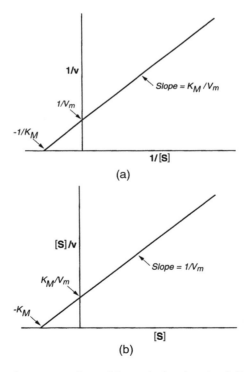

FIGURE 5-2. Linear plots commonly used for analyzing data that follow Michaelis–Menten kinetics. (a) Plot of the reciprocal initial velocity, v, versus the reciprocal substrate concentration, $1/[S]$. (b) Plot of $[S]/v$ versus $[S]$.

This equation predicts that a plot of [S]/v versus [S] should be linear and gives a better weighting of the data than the double reciprocal plot (Eq. 5-16). Obviously, the final results should be independent of how the data are plotted and analyzed! Examples of both plots are given in Figure 5-2.

The steady-state analysis can be extended to much more complex enzyme mechanisms, as well as to other biological processes, but this is beyond the scope of this presentation. How can enzyme catalysis be understood in terms of transition state theory? The simple explanation is that enzyme catalysis lowers the standard free energy of activation. This can be either an entropic or enthalpic effect. Much more detailed interpretations of enzyme catalysis in terms of transition state theory have been developed. The basic idea is that formation of the enzyme–substrate complex alters the free energy profile of the reaction such that the free energy of activation for the reaction is greatly lowered. Essentially, the free energy change associated with binding of the substrate to the enzyme is used to promote catalysis.

5.3 α-CHYMOTRYPSIN

As an example of how mechanisms can be developed for the action of enzymes, early studies of the enzyme α-chymotrypsin will be discussed. Proteolytic enzymes were among the first to be studied, not because of their intrinsic interest, but because they could easily be isolated in a reasonably pure form. In fact, for many years the availability of large quantities of pure enzyme severely restricted the range of enzymes that could be studied in mechanistic detail. With the ease of cloning and modern expression systems, this is no longer a limitation.

The enzyme α-chymotrypsin has a molecular weight of about 25,000 and consists of three polypeptide chains covalently linked by disulfides. It is an *endopeptidase*; that is, it cleaves peptide bonds in the middle of a protein. Enzymes that cleave peptide bonds at the end of a protein are called *exopeptidases*. The overall reaction can be written as

$$R-\overset{\overset{\displaystyle H}{|}}{\underset{\underset{\displaystyle NH}{|}}{C}}-\overset{\overset{\displaystyle O}{\|}}{C}-\overset{\overset{\displaystyle H}{|}}{N}\text{\large\wwww} \quad \xrightarrow{\ H_2O\ } \quad R-\overset{\overset{\displaystyle H}{|}}{\underset{\underset{\displaystyle NH}{|}}{C}}-\overset{\overset{\displaystyle O}{\|}}{C}-O^-\ \ ^+H_3N\text{\large\wwww} \qquad (5\text{-}18)$$

Experimentally, it is found that the enzyme has a strong preference for R being an aromatic group, amino acids phenylalanine, tyrosine, and tryptophan, but hydrophobic amino acids such as isoleucine, leucine, and valine are also good substrates. Studying this reaction with a protein as substrate is very difficult because the molecular structure of the substrate is changing continuously as peptide bonds are cleaved. This makes quantitative interpretation of the kinetics virtually impossible. Consequently, the first step in elucidating the molecular details of the enzyme was to develop "model" substrates, molecules that have the important features of the protein substrate but are much simpler. Chymotrypsin is also a good esterase; that is, it hydrolyzes esters. This proved very useful in elucidating the mechanism of action of the enzyme. The most successful model substrates have the general structure

$$
\begin{array}{c}
\text{H} \quad \text{O} \\
| \quad || \\
\text{R}-\text{C}-\text{C}-\text{NH}_2 \quad \text{Amide} \\
| \\
\text{NH} \\
| \\
\text{C}=\text{O} \\
| \\
\text{CH}_3
\end{array}
\qquad
\begin{array}{c}
\text{H} \quad \text{O} \\
| \quad || \\
\text{R}-\text{C}-\text{C}-\text{O}-\text{R}' \quad \text{Ester} \\
| \\
\text{NH} \\
| \\
\text{C}=\text{O} \\
| \\
\text{CH}_3
\end{array}
$$

where R is the aromatic residue associated with phenylalanine, tyrosine, and tryptophan. Note that the amino group of the aromatic amino acid is acetylated. The enzyme will not work on a free amine, as might be expected for an endopeptidase. Furthermore, only L-amino acids are substrates.

We will consider some of the results obtained with tryptophan as the amino acid.

$$ R = $$

The hydrolysis of tryptophan model substrates follows Michaelis–Menten kinetics, and some of the results obtained for various esters and the amide are shown in Table 5-1. In this table, $k_{cat} = V_m/[E_0]$, and R' is the group covalently linked to the tryptophan carboxyl. The fact that k_{cat} is essentially the same for all esters suggests that a common intermediate is formed and that the breakdown of the intermediate is the rate determining step in the mechanism. A different mechanism must occur for the amide, or there is a different slow step in the mechanism.

An inherent deficiency of steady-state kinetic studies of enzymes is that the enzyme concentration is very low. Consequently, the intermediates in the reaction sequence cannot be detected directly. Limited information about the intermediates can be obtained through steady-state kinetic studies by methods not discussed here, for example, the pH dependence of the kinetic parameters and the use of isotopically labeled substrates that alter the kinetic parameters. However, in order to study the intermediates directly, it is necessary to use sufficiently high enzyme concentrations so that the intermediates can be observed directly. This is the realm of *transient kinetics*. The difficulty for enzymatic reactions is that the reactions become very fast, typically occurring in the millisecond and microsecond range. This requires special experimental techniques. Transient kinetic studies have proved invaluable in elucidating how enzymes work.

In the case of chymotrypsin, a substrate was sought for which a color change occurred upon reaction in order that the changes in concentration could easily and rapidly be observed. The first such substrate studied was *p*-nitrophenyl acetate, which is hydrolyzed by chymotrypsin:

TABLE 5-1. Steady-State Constants for Chymotrypsin

R′	k_{cat} (s^{-1})	K_M (mM)
Ethyl	27	~5
Methyl	28	~5
p-Nitrophenyl	31	~5
Amide	0.03	~0.09

Sources: Adapted from R. J. Foster and C. Niemann, *J. Am. Chem. Soc.*, **77**, 1886 (1955) and L. W. Cunningham and C. S. Brown, *J. Biol. Chem.* **221**, 287 (1956).

$$(5\text{-}19)$$

The advantage of this substrate is that the phenolate ion product is yellow so that the time course of the reaction can easily be followed spectrophotometrically. With this substrate, it was possible to look at the establishment of the steady state, as well as the steady-state reaction. A very important finding was that the reaction becomes extremely slow at low pH. An intermediate, in fact, can be isolated by using a radio-active acetyl group in the substrate. The intermediate is an acylated enzyme. Later studies showed that the acetyl group is attached to a serine residue on chymotrypsin (4).

Based on the kinetic studies and isolation of the acyl enzyme, a possible mechanism is the binding of substrate, acylation of enzyme, and deacylation of the enzyme. The elementary steps can be written as

$$E + S \underset{k_{-1}}{\overset{k_1}{\rightleftharpoons}} ES \overset{k_2}{\rightarrow} E\text{-acyl} \overset{k_3}{\rightarrow} E + Acetate \qquad (5\text{-}20)$$

$$+ p\text{-Nitrophenolate}$$

Does this mechanism give the correct steady-state rate law? The rate law can be derived as for the simple Michaelis–Menten mechanism using the steady-state approximation for all of the enzyme species and the mass conservation relationship for enzyme:

$$[E_0] = [E] + [ES] + [E\text{-acyl}] \qquad (5\text{-}21)$$

$$-\frac{d[ES]}{dt} = (k_{-1} + k_2)[ES] - k_1[E][S] \approx 0$$

$$-\frac{d[E\text{-acyl}]}{dt} = k_3[E\text{-acyl}] - k_2[ES] \approx 0$$

$$v = k_3 [E\text{-acyl}] \tag{5-22}$$

When these relationship are combined, it is found that the Michaelis–Menten rate law, Eq. 5-8, is obtained, with

$$V_m = \frac{k_2 k_3 [E_0]}{k_2 + k_3} \tag{5-23}$$

$$K_M = \left(\frac{k_{-1} + k_2}{k_1}\right)\left(\frac{k_3}{k_2 + k_3}\right) \tag{5-24}$$

For ester substrates, the slow step is deacylation of the enzyme, or $k_3 \ll k_2$. In this case, $k_{cat} = k_3$ and $K_M = [(k_{-1} + k_2)/k_1](k_3/k_2)$.

How did the transient kinetic studies contribute to the postulation of this mechanism? It is worth spending some time to analyze the mechanism in Eq. 5-20 in terms of transient kinetics. This analysis can serve as a prototype for understanding how transient kinetics can be used to probe enzyme mechanisms, not only for chymotrypsin but for other enzyme reactions as well. The proposed mechanism predicts that the acyl enzyme should accumulate and be directly observable. If we consider only a single turnover of the enzyme at very early times and high substrate concentrations, all of the enzyme should be converted to the acyl enzyme, followed by a very slow conversion of the acyl enzyme to the free enzyme through hydrolysis ($k_2 \gg k_3$). The slow conversion to enzyme will follow Michaelis–Menten kinetics (Eq. 5-13) initially as long as the total substrate concentration is much greater than the total enzyme concentration. Thus, the complete solution to the rate equations for the mechanism in Eq. 5-20 should contain two terms: (1) the rate of conversion of the enzyme to the acyl enzyme, and (2) the rate of the overall reaction that is limited by the rate of hydrolysis of the acyl enzyme.

To simplify the analysis, let us assume the following: k_3 is approximately zero at early times ($k_2 \gg k_3$); the formation of the initial enzyme–substrate complex, ES, is very rapid relative to the rates of the acylation and deacylation steps so that ES is in a steady state ($d[ES]/dt = 0$); and the total substrate concentration is much greater than the total enzyme concentration. The last assumption is similar to that for the steady-state approximation used earlier for the Michaelis–Menten mechanism. However, in this case the enzyme concentration is sufficiently high so that the concentration of the intermediate can be detected. The rate of formation of the intermediate is given by

$$\frac{d[\text{E-acyl}]}{dt} = k_2[\text{ES}] \tag{5-25}$$

Because the phenolate ion is formed when the acyl enzyme is formed, the rate of formation of phenolate ion is the same as the rate of formation of the acyl enzyme. This is actually a measurement of the rate of establishment of the steady state. In order to integrate this equation, we make use of the identity

$$[\text{E}] + [\text{ES}] = [\text{ES}](1 + [\text{E}]/[\text{ES}]) = [\text{ES}]\{1 + (k_{-1} + k_2)/(k_1[\text{S}])\}$$

and mass conservation

$$[\text{E}_0] = [\text{E}] + [\text{ES}] + [\text{E-acyl}]$$

or

$$[\text{E}] + [\text{ES}] = [\text{E}_0] - [\text{E-acyl}]$$

Insertion of these relationships into Eq. 5-25 gives

$$\frac{d[\text{E-acyl}]}{dt} = \frac{k_2([\text{E}_0] - [\text{E-acyl}])}{1 + (k_{-1} + k_2)/(k_1[\text{S}])} \tag{5-26}$$

Integration of this equation gives

$$[\text{E-acyl}] = [\text{E}_0](1 - e^{-k_2' t}) \tag{5-27}$$

with $k_2' = k_2/\{1 + (k_{-1} + k_2)/(k_1[\text{S}])\}$.

Note that when $t = 0$, $[\text{E-acyl}] = 0$, and when $t = \infty$, $[\text{E-acyl}] = [\text{E}_0]$, as expected. The total rate of phenolate, P_1, formation at early times is

$$[\text{P}_1] = vt + [\text{E}_0](1 - e^{-k_2' t}) \tag{5-28}$$

The time course of the reaction as described by Eq. 5-28 is shown schematically in Figure 5-3. The second term in the equation dominates at early times, and the exponential rise can be analyzed to give k_2'. The linear portion of the curve corresponds to the Michaelis–Menten initial velocity and can be used to obtain V_m and K_M. The maximum velocity for this limiting case is $k_3[\text{E}_0]$, so that both k_2 and k_3 can be determined. A more exact analysis that does not assume $k_3 = 0$ is a bit more complex, but the end result is similar. The rate equation contains the same two terms as in Eq. 5-28. However, the second term is now more complex: the exponent is $(k_2' + k_3)t$ instead of $k_2' t$; and the amplitude is $[\text{E}_0][k_2'/(k_2' + k_3)]^2$ rather than $[\text{E}_0]$. The first term is the same, vt, with the maximum velocity and Michaelis constant defined by Eqs. 5-23 and 5-24.

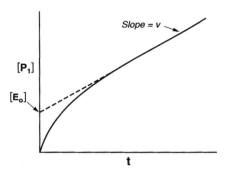

FIGURE 5-3. Schematic representation for the kinetics of an enzymatic reaction displaying "burst" kinetics. The product of the reaction, P_1, is plotted versus the time. As described by Eq. 5-28 for the limiting case of $k_2 \gg k_3$, the slope of the linear portion of the curve is the steady-state initial velocity, v, and the initial exponential time dependence is characterized by the rate constant k_2' and a "burst" amplitude of $[E_0]$.

Finally, we consider the situation when the rate of hydrolysis of the acyl enzyme is very fast relative to its formation ($k_3 \gg k_2$). We have seen previously that fast steps occurring after the rate determining step do not enter into the rate law. Therefore, the mechanism is equivalent to the Michaelis–Menten mechanism with only a single intermediate (Eq. 5-2), with the rate determining step being the acylation of the enzyme ($V_m = k_2[E_0]$). If the transient kinetics are analyzed, the rate is still given by Eq. 5-28, but the amplitude of the second term is essentially zero, rather than $[E_0]$. Thus, the transient kinetics and steady-state kinetics provide the same information. The presence of an initial "burst" of product, as shown in Figure 5-3, is commonly used as a diagnostic for the presence of an intermediate. If the burst is not present, that is, the straight line portion of the curve extrapolates through zero, this means either that an intermediate is not formed, or that its rate of disappearance is much faster than its rate of appearance. This exhaustive, perhaps exhausting, discussion of the mechanism in Eq. 5-20 is a good example of what must be done in order to arrive at a mechanism for an enzyme reaction by kinetic studies. Both steady-state kinetics and transient kinetics are necessary for a complete picture.

The proposed mechanism accounts for all of the facts presented: the observed rate law, the same k_{cat} for all ester substrates, and a common intermediate. What about amides? For amides, it turns out that the slow step is acylation of the enzyme, that is, $k_2 \ll k_3$, so that $k_{cat} = k_2$ and $K_M = (k_{-1} + k_2)/k_1$. As discussed above, if the acylation of the enzyme is slow relative to deacylation, $k_3 \gg k_2$, the intermediate does not accumulate and therefore cannot be observed directly. Literally hundreds of experiments are consistent with this mechanism—a very remarkable achievement. Additional experiments that permitted the very fast reactions prior to the acylation of the enzyme to be studied show that this mechanism is too simple. At least one additional elementary step is required. The sequential elementary steps are: (1) binding of substrate to the enzyme, (2) a conformational change of the enzyme–substrate complex, (3) acylation of the enzyme, and (4) deacylation of the enzyme.

The ultimate goal of kinetic studies is to understand the mechanism in terms of molecular structure. In the case of chymotrypsin, the three-dimensional structure of the enzyme is known and is shown in Color Plate IVa. A well-defined binding pocket is observed for the aromatic side chain of the specific substrates of the enzyme. The pocket is lined with nonpolar side chains of amino acids and is very hydrophobic. Up to three amino acids coupled to the N-terminus of the aromatic amino acid interact with a short range of antiparallel β-sheet in the enzyme. A hydrophobic site also is observed for the amino acid attached to the C-terminus of the substrate. This explains why a free carboxyl group cannot bind and therefore why chymotrypsin is an endopeptidase. As shown in Color Plate IVb, a triad of amino acids, serine-195, histidine-57, and aspartate-102, is observed in the active site region and is found in other serine proteases. The imidazole acts as a general base during the nucleophilic attack of the serine hydroxyl on the substrate. A tetrahedral intermediate probably is formed prior to acyl enzyme formation. Water serves as the nucleophile for the hydrolysis of the acyl enzyme, with imidazole again participating as the general base. An abbreviated version of the catalytic mechanism is shown in Figure 5-4.

This greatly truncated story of the elucidation of the mechanism of action of chymotrypsin illustrates several important concepts: the steady-state approximation; the use of steady-state kinetics in determining chemical mechanisms; the importance of transient kinetics in elucidating intermediates in a mechanism; and the interpretation of the mechanism in terms of molecular structure. More complete discussions of chymotrypsin are available (1,2,5).

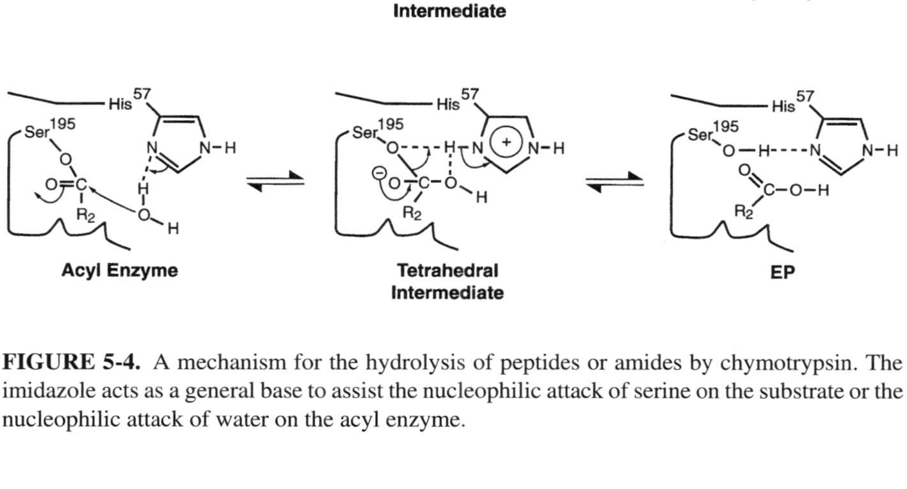

FIGURE 5-4. A mechanism for the hydrolysis of peptides or amides by chymotrypsin. The imidazole acts as a general base to assist the nucleophilic attack of serine on the substrate or the nucleophilic attack of water on the acyl enzyme.

5.4 PROTEIN TYROSINE PHOSPHATASE

In our earlier discussion of metabolism, we saw that the energy obtained in glycolysis is stored as a phosphate ester in ATP and serves as a source of free energy for biosynthesis. The importance of phosphate esters is not confined to this function. Phosphorylation and dephosphorylation of proteins plays a key role in signal transduction and the regulation of many cell functions. For example, the binding of hormones to cell surfaces can trigger a cascade of such reactions that regulate metabolism within the cell. The cell processes modulated by this mechanism include T-cell activation, the cell cycle, DNA replication, transcription and translation, and programmed cell death, among many others. Both the kinases, responsible for protein phosphorylation, and the phosphatases, responsible for protein dephosphorylation, have been studied extensively. We shall restrict this discussion to the phosphatases, and primarily to one particular type of phosphatase.

The phosphorylation of proteins primarily occurs at three sites, namely, the side chains of threonine, serine, and tyrosine. Protein phosphatases can be divided into two structural classes: the serine/threonine specific phosphatases that require metal ions; and tyrosine phosphatases that do not require metals but use a nucleophilic cysteine to cleave the phosphate. The latter class includes dual function phosphatases that can hydrolyze phosphate esters on serine and threonine, as well as tyrosine. Many reviews of these enzymes are available (cf. Refs. 6 and 7). We will discuss the mechanism of a particular example of tyrosine specific phosphatase, namely, an enzyme found in both human and rat that has been extensively studied structurally and mechanistically (6).

The overall three-dimensional structure of the catalytic domain of the enzyme is shown in Color Plate Va, and the active site region is shown in Color Plate Vb with vanadate bound to the catalytic site. Vanadate is frequently used as a model for phosphate esters because the oxygen bonding to vanadium is similar to the oxygen bonding to phosphorus, and vanadate binds tightly to the enzyme. A cysteine rests at the base of the active-site cleft. An aspartic acid and threonine (or serine in similar enzymes) are also near the active site and are postulated to play a role in catalysis, as discussed below.

The kinetic mechanism that has been postulated is the binding of the protein substrate followed by transfer of the phosphoryl group from tyrosine to the cysteine at the active site, and finally hydrolysis of the phosphorylated cysteine:

$$E + S \underset{k_{-1}}{\overset{k_1}{\rightleftharpoons}} ES \overset{k_2}{\longrightarrow} E\text{-}PO_3 \overset{k_3}{\longrightarrow} E + HPO_4 \qquad (5\text{-}29)$$

As with chymotrypsin, a protein is not a convenient substrate to use because of the difficulty in phosphorylating a single specific tyrosine and the lack of a convenient method for following the reaction progress. In this case the model substrate used was p-nitrophenyl phosphate. Similar to chymotrypsin, when the enzyme is phosphorylated the yellow phenolate ion is released so that the reaction progress can easily be monitored. Extensive steady-state and transient kinetic studies have been carried out, and we will only present a few selected results that summarize the findings (8,9).

The role of the cysteine was firmly established by site specific mutagenesis, as converting the cysteine to a serine results in an inactive enzyme. The cysteine is postulated to function as a nucleophile with the formation of a cysteine phosphate. However, when transient kinetic studies were carried out with the native enzyme, a burst phase was not observed. This is shown in Figure 5-5a. As we noted previously, this indicates that either a phosphoenzyme intermediate is not formed, or its hydrolysis is much faster than its formation. The aspartic acid and serine or threonine shown at the active site of the enzyme in Color Plate V are conserved in many different enzymes of this class of tyrosine phosphatases. Consequently, it was decided to mutate these residues. If the serine is mutated to alanine, the enzyme is still active, but now a burst is observed, as shown in Figure 5-5b. This burst was postulated to represent the formation of the phosphoenzyme, and the rate constants were obtained for the phosphorylation of cysteine and its hydrolysis. The results of the transient kinetic experiments are presented in Table 5-2. For the native enzyme, only k_{cat} can be determined and is equal to k_2. For the serine to alan-

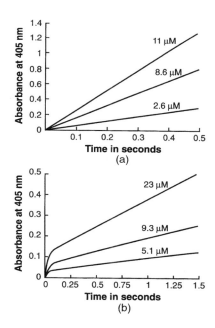

FIGURE 5-5. Transient kinetic time course for the hydrolysis of p-nitrophenyl phosphate by native (a) and mutant (b) protein tyrosine phosphatase. The point mutation substituted alanine for serine 222. The concentrations of substrate are shown next to the traces. Reprinted with permission from D. L. Lohse, J. M. Denu, N. Santoro, and J. E. Dixon, *Biochemistry* **36**, 4568 (1997). Copyright © 1997 American Chemical Society.

TABLE 5-2. **Kinetic Parameters for Native and Mutated Tyrosine Phosphatases**

Enzyme	k_{cat} (s^{-1})	k_2 (s^{-1})	k_3 (s^{-1})	K (μM)^a
Native	20			
Serine 222 → Alanine	1.3	34	1.4	0.46
Aspartate 181 → Asparagine	0.27			
Serine 222 → Alanine and	0.055	1.9	0.057	0.81
Aspartate 181 → Asparagine				

Source: Adapted from D. L. Lohse, J. M. Denu, N. Santoro, and J. E. Dixon, *Biochemistry* **36**, 4568 (1997).
$^a(k_{-1} + k_2)/k_1$.

ine mutant, the rate constant for hydrolysis of the intermediate is the same as k_{cat}, as expected. This mutation also suggests that serine plays a role in the hydrolysis mechanism.

The importance of the aspartic acid in the mechanism was inferred not only from its conservation in many different enzymes, but also by the pH dependence of the steady-state kinetic parameters, which suggested it functioned as a general acid in the phosphorylation of the enzyme. When the aspartic acid was mutated to asparagine, the enzyme still functioned, but k_{cat} was only about 1% of that of the native enzyme (Table 5-2). The transient kinetics did not show a burst phase, indicating that formation of the phosphoenzyme was still rate determining. As expected, the pH dependence of the steady-state kinetic parameters was altered by this mutation.

Finally, the double mutated enzyme was prepared in which the serine was changed to alanine and the aspartate to asparagine. The transient kinetics showed a burst so that the slow step was now the hydrolysis of the intermediate, as with the single mutant in which serine was changed to alanine. However, k_{cat} was reduced even further, and the rate constants for both phosphorylation of the enzyme and hydrolysis were greatly reduced (Table 5-2). On the basis of these and other data, the aspartate was postulated to serve as a general base in the hydrolysis of the intermediate. The intermediate is sufficiently stable in the double mutant so that it could be observed directly with phosphorus nuclear magnetic resonance (8). The mechanism proposed for this reaction has two distinct transition states, one for phosphorylation of the cysteine and another for hydrolysis of the phosphorylated enzyme. Based on the experiments discussed and other data, especially structural data, structures have been proposed for the two transition states as shown in Figure 5-6.

The mechanism of tyrosine phosphatases illustrates several important points. First, the general usefulness of the analysis developed for chymotrypsin is apparent. Both transient and steady-state kinetic experiments were important in postulating a mechanism. Second, the importance of site specific mutations in helping to establish the mechanism is evident. This method is not without pitfalls, however. It is important to establish that mutations do not alter activity through structural changes in the molecule. In the present case, experiments were done to establish the structural integrity of the mutant enzymes. This ensures that the mutations are altering only the chemical aspects of the mechanism. Finally, this dis-

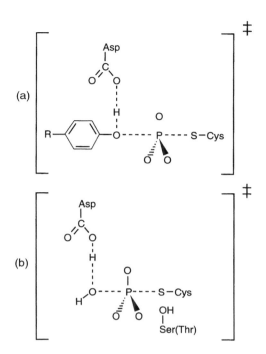

FIGURE 5-6. Proposed transition states for the protein tyrosine phosphatase reaction. (a) Formation of the cysteine phosphate intermediate with a tyrosine phosphate as substrate. (b) Hydrolysis of the cysteine phosphate intermediate. Adapted from J. M. Denu, D. L. Lohse, J. Vijayalakshmi, M. A. Saper, and J. E. Dixon, *Proc. Natl Acad. Sci. USA* **93**, 2493 (1996). Republished with permission of the National Academy of Sciences USA. Reproduced by permission of the publisher via Copyright Clearance Center, Inc.

cussion again illustrates the many different types of experiments that must be done in developing mechanisms. The kinetic results must be bolstered by structural and chemical information. Many other studies of enzymes could profitably be discussed, but instead we will turn our attention to kinetic studies in other types of biological systems.

5.5 RIBOZYMES

Enzymes are the most efficient and prevalent catalysts in physiological systems, but they are not the only catalysts of biological importance. RNA molecules have also been found to catalyze a wide range of reactions. Most of these reactions involve the processing of RNA, cutting RNA to the appropriate size or splicing RNA. RNA has also been implicated in peptide bond formation on the ribosome and has been shown to hydrolyze amino acid esters. These catalytic RNAs are called *ribozymes*. They are much less efficient than a typical protein enzyme and sometimes catalyze only a single event. In the latter case, this is not true catalysis. Ribozymes also may work in close collaboration with a protein in the catalytic event. We will consider an abbreviated discussion of a ribozyme, ribonuclease P. For more information, many reviews of ribozymes are available (10–12).

Ribonuclease P catalyzes an essential step in tRNA maturation, namely, the cleavage of the 5′ end of a precursor tRNA (pre-tRNA) to give an RNA fragment and the mature tRNA. This reaction is shown schematically in Figure 5-7. Also shown in this figure is a representation of the catalytic RNA. The catalytic RNA is about 400 nucleotides long and catalyzes the maturation reaction by itself *in vitro*. *In vivo*, a protein of about 120 amino acids also participates in the catalysis. The precise role of the protein is not known, but it appears to alter the conformation of the RNA. The protein–RNA enzyme has a broader selectivity for biological substrates and is a more efficient catalyst (13). Metal ions are also involved in this reaction, with Mg^{2+} being the most

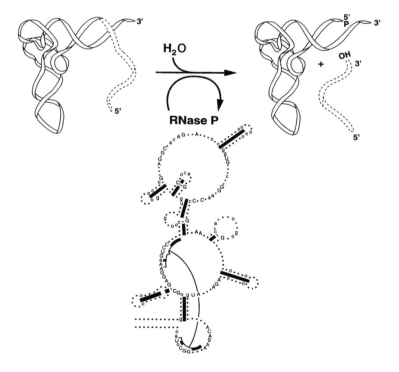

FIGURE 5-7. The reaction of pre-tRNA catalyzed by ribonuclease P is shown schematically at the top of the figure. The 5′ leader sequence of the pre-tRNA is removed in the reaction. The eubacterial consensus structure of ribonuclease P RNA is shown in the lower part of the figure. Proven helices are designated by filled rectangles, invariant nucleotides by uppercase letters, >90% conserved nucleotides by lowercase letters, and less conserved nucleotides by dots. Reproduced from J. W. Brown and N. R. Pace, *Nucleic Acids Res.* **20**, 1451 (1992). Reproduced by permission of Oxford University Press.

important physiologically. Metal ions probably play a role in both catalysis and in maintaining the active conformation of the RNA, but we will not consider this aspect of the reaction here. Transient kinetic studies of the ribozyme have been carried out, and the minimal mechanism consists of the binding of pre-tRNA to the ribozyme, cleavage of the phosphodiester bond, and independent dissociation of both products. We will present some of the results obtained with the RNA component of ribonuclease P.

If transient experiments are carried out with ribonuclease P with excess substrate, a "burst" of the tRNA product occurs at short times, followed by an increase in the product concentration that is linear with time. This behavior is familiar by now, as it has been observed for chymotrypsin and protein tyrosine phosphatase, as discussed above. In the case of ribonuclease P, a covalent intermediate is not formed with the ribozyme; instead, the dissociation of products is very slow relative to the hydrolytic reaction. The mechanism can be written as

$$\text{pre-tRNA} + \text{E} \rightleftharpoons \text{ES} \rightarrow \text{tRNA} + \text{P} + \text{E} \tag{5-30}$$

where E is RNAase P RNA and P is the pre-tRNA fragment product. This is not exactly the same as the mechanism discussed for chymotrypsin, but the analysis is similar if the second step is assumed to be much slower than the first. (The detailed analysis is given in Ref. 14.) There is an important lesson to be learned here. "Burst" kinetics are observed whenever the product is formed in a rapid first step followed by regeneration of the enzyme in a slower second step. The apparent rate constants for these two steps can be determined.

What is desired, however, is to distinguish between the binding of enzyme and substrate and the hydrolysis; both of these reactions are aggregated into the first step in Eq. 5-30. In the case of chymotrypsin and protein tyrosine phosphatase, we assumed that the binding step was rapid relative to the formation of the enzyme–substrate intermediate. In this case, we cannot assume that binding is rapid relative to hydrolysis. The two steps were resolved by single turnover experiments: an excess of enzyme was used and because dissociation of products is slow, only a single turnover of substrate was observed (14). The mechanism can be written as

$$\text{E} + \text{pre-tRNA} \xrightarrow{k_1} \text{E·pre-tRNA} \xrightarrow{k_2} \text{E·tRNA·P} \tag{5-31}$$

The results obtained are shown in Figure 5-8 for two different concentrations of enzyme. At the lower concentration of enzyme, it is clear that the curve is sigmoidal, rather than hyperbolic. This is because the first step is sufficiently slow that it takes some time for the concentration of the first enzyme–substrate complex to build up. The data at the higher enzyme concentration are essentially hyperbolic because the rate of the first reaction is greater at higher enzyme concentration. These data can be explained by analyzing the mechanism in Eq. 5-30 as two consecutive irreversible first order reactions. Why irreversible? Fortunately, the rate of dissociation of pre-tRNA

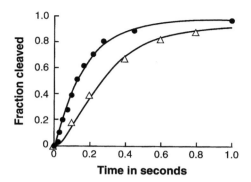

FIGURE 5-8. Single turnover measurements of the hydrolysis of pre-tRNA catalyzed by the RNA component of RNAase P. The fraction of pre-tRNA cleaved is plotted versus time. The pre-tRNA was mixed with excess concentrations of RNAase P RNA, 1.4 µM ($\triangle$) or 19 µM ($\bigcirc$). The data are the fit to a mechanism of two consecutive first order reactions (Eq. 5-38). Reprinted with permission from J. A. Beebe and C. A. Fierke, *Biochemistry* **33**, 10294 (1994). Copyright © 1994 American Chemical Society.

from the ribozyme is sufficiently slow that it does not occur on the time scale of the experiment, and furthermore, the analysis of the product gives both E-tRNA and free tRNA so that if some dissociation of tRNA occurs it is not relevant. Why is the first step assumed to be first order? This is another example of a pseudo first order rate constant because the enzyme concentration is effectively constant throughout the experiment. The pseudo first order rate constant is $k_1[E]$, where k_1 is the second order rate constant for the reaction of the enzyme with the substrate.

To simplify the nomenclature, we will rewrite the mechanism as

$$A \rightarrow B \rightarrow C \tag{5-32}$$

The rate equation for the time dependence of A is

$$-\frac{d[A]}{dt} = k_1[A] \tag{5-33}$$

which is easily integrated to give

$$[A] = [A]_0 e^{-k_1 t} \tag{5-34}$$

where $[A]_0$ is the starting concentration, in this case of pre-tRNA. The rate equation for the time dependence of B is

$$-\frac{d[B]}{dt} = k_2[B] - k_1[A] \tag{5-35}$$

$$-\frac{d[B]}{dt} = k_2[B] - k_1[A]_0 e^{-k_1 t}$$

The solution to this differential equation is

$$[B] = \frac{[A]_0 k_1}{k_2 - k_1}(e^{-k_1 t} - e^{-k_2 t}) \qquad (5\text{-}36)$$

This solution does not work if $k_1 = k_2$ as the denominator goes to zero. For this special case,

$$[B] = k_1[A]_0 t e^{-k_1 t} \qquad (5\text{-}37)$$

Finally, the time dependence of C can be obtained from mass balance since $[A]_0 = [A] + [B] + [C]$:

$$[C] = [A_0]\left(1 - \frac{k_2}{k_2 - k_1}e^{-k_1 t} + \frac{k_1}{k_2 - k_1}e^{-k_2 t}\right) \qquad (5\text{-}38)$$

If Eq. 5-38 is used to analyze the data in Figure 5-8, it is found that $k_1 = 6 \times 10^6$ M^{-1} s^{-1} [RNA]$_0$ and $k_2 = 6$ s^{-1}.

Understanding how to analyze "burst" kinetics and consecutive first order reactions is sufficient for the kinetic analysis of many enzymatic reactions, as the conditions can usually be adjusted to conform to these relatively simple mechanisms.

The mechanistic work carried out with RNAseP has permitted the establishment of a minimal mechanism for the RNA portion of the enzyme. However, much remains to be done. The roles of metals, specific groups on the RNA, and the protein remain to be delineated. Understanding how ribozymes function is at the forefront of modern biochemistry and has important implications for both physiology and the evolution of enzymes.

5.6 DNA MELTING AND RENATURATION

We will conclude this chapter with a discussion of the denaturation and renaturation of DNA. Understanding the dynamics of such processes is clearly of biological importance. At the outset, it must be stated that a detailed understanding of the kinetics and mechanisms have not been achieved. However, this is not for lack of effort, and a qualitative understanding of the mechanisms has been obtained.

We will begin with some of the elementary steps in the dynamics of the interactions between the two chains of helical DNA. As discussed in Chapter 3, the thermodynamics of hydrogen bonding between bases has been studied in nonaqueous solvents, where the dimers formed are reasonably stable. Kinetic studies of hydrogen bonded dimers also have been carried out, and the reactions have been found to be extremely fast, occurring on the submicrosecond and nanosecond time scale. For example, the kinetics of formation of a hydrogen bonded dimer between 1-cyclohexyluracil and 9-ethyladenine has been studied in chloroform (15). (The hydrocarbon arms have been added to the bases to increase their solubility in chloroform.) The reaction is found to occur in a single step with a second order rate constant for the formation of the dimer of $2.8 \times 10^9 \ M^{-1} \ s^{-1}$ and a dissociation rate constant of $2.2 \times 10^7 \ s^{-1}$ at 20°C. The second order rate constant is the maximum possible value; that is, it is the value expected if every collision between the reactants produced a hydrogen bonded dimer. The upper limit for the rate constant of a bimolecular reaction can be calculated from the known rates of diffusion of the reactants in the solvent. In all cases where hydrogen bonded dimer formation has been studied, the formation of the dimer has been found to be diffusion controlled.

What does this tell us about the rate of hydrogen bond formation that occurs after the two reactants have diffused together? To answer this question, we will postulate a very simple mechanism, namely, diffusion together of the reactants to form a dimer that is not hydrogen bonded, followed by the formation of hydrogen bonds. This can be written as

$$A + B \; \underset{k_{-1}}{\overset{k_1}{\rightleftharpoons}} \; [A,B] \; \underset{k_{-2}}{\overset{k_2}{\rightleftharpoons}} \; A\text{-}B \tag{5-39}$$

In this mechanism, k_1 is the rate constant for the diffusion controlled formation of the intermediate, and k_{-1} is the rate constant for the diffusion controlled dissociation of the intermediate. Since only a single step is observed in the experiments, assume that the initial complex formed is in a steady state:

$$[A,B] = k_1[A][B]/(k_{-1} + k_2) + k_{-2}[A\text{-}B]/(k_{-1} + k_2) \tag{5-40}$$

The overall rate of the reaction is

$$\frac{d[A\text{-}B]}{dt} = k_2[A,B] - k_{-2}[A\text{-}B] \tag{5-41}$$

If Eq. 5-40 is substituted into Eq. 5-41, we obtain

$$\frac{d[A\text{-}B]}{dt} = k_f[A][B] - k_r[A\text{-}B] \tag{5-42}$$

with

$$k_f = k_1 k_2/(k_{-1} + k_2)$$

and

$$k_r = k_{-1} k_{-2}/(k_{-1} + k_2)$$

The experimental results indicate that $k_f = k_1$. This is true if $k_2 \gg k_{-1}$, or formation of the hydrogen bonds is much faster than diffusion apart of the intermediate. However, we can calculate the value of k_{-1}; it is about 10^{10} s^{-1}. Therefore, the rate constant for formation of the hydrogen bonds, k_2, must be greater than about 10^{11} s^{-1}. A more exact analysis would put in a separate step for formation of each of the two hydrogen bonds, but this would not change the conclusion with regard to the rate of hydrogen bond formation. Hydrogen bond formation is very fast!

We know that hydrogen bonding alone cannot account for the stability of the DNA double helix: Base stacking and hydrophobic interactions also are important. An estimate of the rate of base stacking formation and dissociation has been obtained in studies of polyA and polydA (16). These molecules undergo a transition from "stacked" to "unstacked" when the temperature is raised. This transition is accompanied by spectral changes and can easily be monitored. Conditions were adjusted so that the molecules were in the middle of the transition, and the temperature was raised very rapidly (<1 μs) with a laser. This is an example of a kinetic study near equilibrium. As the system returns to equilibrium at the higher temperature, the relaxation time can be measured and is the sum of the rate constants for stacking and unstacking (Eq. 4-43). The results obtained indicate that the rate constants are in the range of $10^6 - 10^7$ s^{-1}. Although this is a very fast process, it is considerably slower than hydrogen bonding or simple rotation of the bases.

Moving up the ladder of complexity, we now consider the formation of hydrogen bonded dimers between oligonucleotides (17,18). Again, only a few prototype reactions will be considered, namely, the reaction of a series of A_n oligonucleotides with U_n oligonucleotides to form double helical dimers. These reactions occur in the millisecond to second time regime with $n = 8–18$. In all cases, only a single relaxation time was observed. The relaxation time was consistent with a simple bimolecular reaction (Eq. 4-47):

$$1/\tau = k_1([A_n]_e + [U_n]_e) + k_{-1} \tag{5-43}$$

A typical plot of $1/\tau$ versus the sum of the equilibrium concentration is given in Figure 5-9, and some of the rate constants obtained are given in Table 5-3. The bimolecular rate constants are all about 10^6 M^{-1} s^{-1}, whereas the dissociation rate constants vary widely, reflecting the stability of the double helix that is formed. Although the second order rate constant is quite large, it is considerably below the value expected for a diffusion controlled reaction.

A detailed analysis of the mechanism can be made as above for the case of the formation of a hydrogen bonded dimer. This case is considerably more complex, as a separate step is needed for the pairing of each of the n bases. The analysis carried out was bolstered by determination of the activation energies for the rate constants. The conclusion reached is that *nucleation* of the double helix after the molecules have diffused together is rate determining. The formation of the first base pair is rapid, but dissociation is even more rapid. This is also true for the formation of two base pairs. However, when the third base pair is formed, the structure is stabilized. After a nucleus

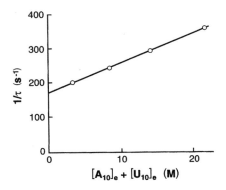

FIGURE 5-9. A plot of $1/\tau$ versus the sum of the equilibrium concentrations of A_{10} and U_{10} at 17°C (Eq. 5-43). The two reactants combine to form a double helix structure. Adapted from D. Pörschke and M. Eigen, *J. Mol. Biol.* **62**, 361 (1971). Reproduced with the permission of Academic Press, Inc.

of three base pairs is present, each later base pair forms with a specific rate constant of about 10^7 s^{-1}. Note that this is consistent with the results obtained for the "stacking" and "unstacking" of polyA. Altogether these results indicate that the elementary steps in the dynamics of DNA base pairing and stacking (and whatever other types of interactions are important for base pairing) are very rapid. We now turn to consideration of DNA itself.

As might be expected, the melting and reformation of DNA is very complex. DNA is very heterogeneous, and we have discussed previously (Chapter 3) that A–T rich regions are much less thermodynamically stable than G–C rich regions. In addition, DNA has a complex structure, with loops and bulges. Native DNA has a very precise alignment of base pairs. If extensive melting occurs, base pairings other than the native pairings may form when the temperature is lowered to form the native structure, thereby slowing down reformation of the native structure. The ultimate form of melting is separation of the individual strands. Because the strands are very long, the rate of melting could be limited by the hydrodynamics of moving the chains apart. Bearing

TABLE 5-3. Rate Constants for the Reaction of $[A_n]$ with $[U_n]$ at 17°C

n	$10^6 k_1$ $(M^{-1}$ $s^{-1})$	k_{-1} (s^{-1})
8	1.7	2400
9	7.1	640
10	0.83	175
11	0.58	28
14	1.44	1

Source: Adapted from D. Pörschke and M. Eigen, *J. Mol. Biol.* **62**, 361 (1971).

these factors in mind, let us examine some experimental results. Fortunately, the melting and reformation can easily be followed because the spectrum of DNA changes when the bases are stacked. Furthermore, the melting/reformation processes are easy to initiate by changing the temperature.

Some typical results for the denaturation of a viral DNA are shown in Figure 5-10. The denaturation was initiated by the raising of the temperature to give a loss in the native spectrum of 25%, 53%, and 95%. In the first case, the loss in absorbance is approximately first order with a rate constant of about 1 min^{-1}. The other two cases do not strictly follow first order kinetics; that is, the logarithmic plot in Figure 5-10 is not linear, and the rates are clearly much slower. The approximate first order rate constants at long times are shown in the figure. In contrast to the results in model systems, the melting kinetics cannot be fit to a single relaxation time, and the time scale is 100–1000 s. The viral DNA studied has about 2×10^5 base pairs. If only the elementary steps discussed above were involved, the DNA would be completely melted in less than 1 second. How can these results be explained? The rate of melting is governed by the rate of unwinding of the double helix, and the rate of unwinding varies as the melting proceeds. Melting will usually start in the interior of the DNA, and unwinding can only occur if one end of the loop rotates with respect to the other. Initially this rotation is fast because the double helix is approximately a rod, but as melting proceeds, the bulky loops produced make it harder for one end of the helix to rotate with respect to the other. The rate then becomes slower, as observed. Thus, strand unwinding and separation become rate limiting. In support of this mechanism, if DNA is subjected to a very drastic denaturing force such as a very high pH, all of the base pairs rapidly break in times less than 0.01 s. This does not involve unwinding and strand separation

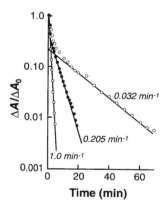

FIGURE 5-10. The kinetics of T2 DNA denaturation in 0.08 M NaCl, 80% formamide. The fraction of the absorbance at 259 nm associated with the native structure is plotted versus the time. The reaction was initiated by heating the three samples to a loss in the native ultraviolet spectrum of 25% (□), 53% (●), and 95% (○). Reproduced from M. T. Record and B. H. Zimm, *Biopolymers* **11**, 1435 (1972). Copyright © 1972 *Biopolymers*. Reprinted by permission of John Wiley & Sons, Inc.

as the native DNA can rapidly be formed by lowering the pH. A detailed description of this mechanism has been developed (19).

The slow unwinding time could be a severe problem for the replication of large DNAs. However, nature has solved this problem by creating enzymes that (1) can make breaks in the DNA chains, thereby reducing the length of DNA that must be rotated during the unwinding, and (2) can unwind DNA.

As a final example, we consider the renaturation of separated DNA strands. DNA can be denatured by raising the temperature significantly above the midpoint of the melting transition, or by raising the pH. After the strands have separated, native DNA can be formed by lowering the temperature or lowering the pH. As might be expected, the renaturation process is second order. The rate of reaction depends on the source of the DNA and its length. The dependence on length can be eliminated by first sonicating the DNA to form fragments of about the same size. If the rates of renaturation are measured at the same DNA concentration in moles of nucleotides per liter, they vary by several orders of magnitude, as shown in Figure 5-11. The figure includes data for polyA–polyU and a double stranded RNA (MS-2 viral RNA). The scale at the top is the number of nucleotide pairs in the genome. The larger the genome, the smaller the number of complementary fragments. Only complementary strands can form native DNA, so that the smaller the number of complementary strands, the slower the rate. For polyU and polyA, every strand is complementary so that this system is assigned a value of 1 on this scale.

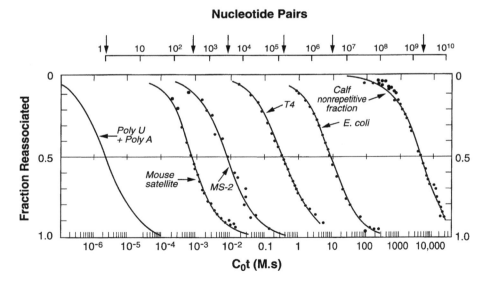

FIGURE 5-11. Reassociation of double-stranded nucleic acids from various sources. The genome size, which is a measure of the complexity of the DNA, is shown above the figure. The nucleic acids all have a single-stranded chain length of about 400 bases. Adapted with permission from R. J. Britten and D. E. Kohne, *Science* **161**, 529 (1968). Copyright © 1968 American Association for the Advancement of Science.

The renaturation process can be analyzed quantitatively. If the complementary strands are represented as A and A′, the reaction can be written as

$$A + A' \xrightarrow{k} AA' \tag{5-44}$$

Since the concentrations of A and A′ are equal, the rate equation is

$$-\frac{d[A]}{dt} = k[A]^2 \tag{5-45}$$

which can be integrated to give

$$1/[A] - 1/[A]_0 = kt \tag{5-46}$$

The data were found to obey this equation. The half-time for the reaction, that is, when $[A] = [A]_0/2$, is given by

$$t_{1/2} = 1/(k[A]_0) \tag{5-47}$$

In the above equations, the concentration should be equal to the concentration of complementary strands, which we generally do not know. If the total number of base pairs in the smallest repeating sequence is N, then the concentration of complementary strands is proportional to the total concentration of DNA, c_0, divided by N. If this relationship is substituted in Eq. 5-47, we find that $c_0 t_{1/2} \propto N$. Thus, the total concentration of DNA times the half-time for renaturation is a measure of N. A long half-time means that a given fragment will have to sample many other fragments before it finds its complement, so the sequence is complex. N is a measure of the "complexity" of the DNA, and the scale on the top of Figure 5-11 can be equated to N. On this basis, the calf DNA is the most complex DNA that was studied. The study of the kinetics of renaturation can be a useful probe of a gene. It has been used, for example, to determine if fragments of a viral gene were present in viral transformed cells (20).

This concludes the sampling of the application of kinetics to biological systems. Many interesting studies have not been discussed, and many more remain to be done. The study of the time dependence of biological processes is essential in the quest for the elucidation of molecular mechanisms.

REFERENCES

1. G. G. Hammes, *Enzyme Catalysis and Regulation*, Academic Press, New York, 1982.

2. A. Fersht, *Structure and Mechanism in Protein Science: A Guide to Enzyme Catalysis and Protein Folding,* W. H. Freeman, San Francisco, 1999.

3. D. L. Purich (ed.), *Contemporary Enzyme Kinetics and Mechanism*, Academic Press, New York, 1996.

4. B. S. Hartley and B. A. Kilby, *Biochem. J.* **56**, 288 (1954).

5. G. P. Hess, in *The Enzymes*, 3rd edition (P. Boyer, ed.), Vol. 3, p. 213, Academic Press, New York, 1970.

6. J. M. Denu, J. A. Stuckey, M. A. Saper, and J. E. Dixon, *Cell* **87**, 361 (1996).

7. S. Shenolikaar, *Annu. Rev. Cell Biol.* **10**, 55 (1994).

8. J. M. Denu, D. L. Lohse, J. Vijayalakshmi, M. A. Saper, and J. E. Dixon, *Proc. Natl. Acad. Sci. USA* **93**, 2493 (1996).

9. D. L. Lohse, J. M. Denu, N. Santoro, and J. E. Dixon, *Biochemistry* **36**, 4568 (1997).

10. T. R. Cech, in *The RNA World*, Cold Spring Harbor Laboratory Press, Cold Spring Harbor, NY, 1993, p. 239.

11. D. B. McKay and J. E. Wedekind, in *The RNA World*, 2nd edition, Cold Spring Harbor Laboratory Press, Cold Spring Harbor, NY, 1999, p. 265.

12. S. Altman and L. Kirsebom, in *The RNA World*, 2nd edition, Cold Spring Harbor Laboratory Press, Cold Spring Harbor, NY, 1999, p. 351.

13. A. Loria, S. Niranjanakumari, C. A. Fierke, and T. Pan, *Biochemistry* **37**, 15466 (1998).

14. J. A. Beebe and C. A. Fierke, *Biochemistry* **33**, 10294 (1994).

15. G. G. Hammes and A. C. Park, *J. Am. Chem. Soc.* **91**, 956 (1969).

16. T. G. Dewey and D. H. Turner, *Biochemistry* **18**, 5757 (1979).

17. D. Riesner and R. Romer, in *Physico-Chemical Properties of Nucleic Acids*, Vol. 2 (J. Duchesne, ed.), Academic Press, London, 1973.

18. D. Pörschke and M. Eigen, *J. Mol. Biol.* **62**, 361 (1971).

19. M. T. Record and B. H. Zimm, *Biopolymers* **11**, 1435 (1972).

20. S. J. Flint, P. H. Gallimore, and P. A, Sharp, *J. Mol. Biol.* **96**, 47 (1975).

PROBLEMS

5-1. The hydration of CO_2 is catalyzed by the enzyme carbonic anhydrase. The overall reaction at neutral pH can be written as

$$CO_2 + H_2O \rightleftharpoons HCO_3^- + H^+$$

The steady-state kinetics of both hydration and dehydration have been studied at pH 7.1, 0.5°C. Some typical data are given below for an enzyme concentration of 2.8×10^{-9} M.

Hydration		Dehydration	
$10^3/v$ (M⁻¹s)	$10^3[CO_2]$ (M)	$10^3/v$ (M⁻¹s)	$10^3[CO_3^-]$ (M)
36	1.25	95	2.0
20	2.5	45	5.0
12	5.0	29	10
6.0	20	25	15

Calculate the steady-state parameters for the forward and reverse reactions.

5-2. Studies of the inhibition of enzymes by various compounds often provide information about the nature of the binding site and the mechanism. Competitive inhibition is when the inhibitor, I, competes with the substrate for the catalytic site. This mechanism can be written as

$$E + S \rightleftharpoons ES \rightarrow E + P$$

$$E + I \rightleftharpoons EI$$

Derive the steady-state rate law for this mechanism and show that it follows Michaelis–Menten kinetics when the inhibitor concentration is constant. Assume the inhibitor concentration is much greater than the enzyme concentration.

5-3. In the text, the steady-state rate law was derived with the assumption that the reaction is irreversible and/or only the initial velocity was determined. Derive the steady-state rate law for the reversible enzyme reaction:

$$E + S \rightleftharpoons X \rightleftharpoons E + P$$

Show that the rate law can be put into the form

$$v = \frac{V_S/K_S[S] - V_P/K_P[P]}{1 + [S]/K_S + [P]/K_P}$$

where V_S and V_P are the maximum velocities for the forward and reverse reactions, and K_S and K_P are the Michaelis constants for the forward and reverse reactions.

When equilibrium is reached, $v = 0$. Calculate the ratio of the equilibrium concentrations of S and P, $[P]/[S]$, in terms of the four steady-state parameters. This relationship is called the Haldane relationship and is a method for determining the equilibrium constant of the overall reaction.

5-4. The kinetics of the formation and breakdown of hydrogen bonded loops or hairpins within small RNA molecules can be studied with relaxation methods since the formation of the helical structures is accompanied by a spectral change. This reaction can be represented as

$$\text{Hydrogen bonded} \underset{k_2}{\overset{k_1}{\rightleftharpoons}} \text{Non-hydrogen bonded}$$

What is the relaxation time for this reaction?

The relaxation time determined at a specific concentration was found to be 10 μs. Calculate the individual rate constants if the equilibrium constant is 0.5.

If the concentration of the RNA is doubled, would the relaxation time get smaller, larger, or stay the same?

5-5. Consider the binding of a protein, P, to a DNA segment (gene regulation). Assume that only one binding site for P exists on the DNA and that the concentration of DNA binding sites is much less than the concentration of P. The reaction mechanism for binding can be represented as

$$P + DNA \rightleftharpoons P\text{-}DNA \rightarrow P\text{-}DNA'$$

where the second step represents a conformational change in the protein. Calculate the rate law for the appearance of P-DNA′ under the following conditions.

A. The first step in the mechanism equilibrates rapidly relative to the rate of the overall reaction and $[P\text{-}DNA] \ll [DNA]$.

B. The intermediate, P-DNA, is in a steady state.

C. The first step in the mechanism equilibrates rapidly relative to the rate of the overall reaction and the concentrations of DNA and P-DNA are comparable. Express the rate law in terms of the *total* concentration of DNA and P-DNA, that is, $[DNA] + [P\text{-}DNA]$.

D. The following initial rates were measured with an initial DNA concentration of 1 μM.

[P] (μM)	10^4 Rate (M/s)
100	8.33
50	7.14
20	5.00
10	3.33

Which of the rate laws is consistent with the data?

5-6. Many biological reactions are very sensitive to pH. This can readily be incorporated into the rate laws because protolytic reactions can be assumed to be much faster than other rates in most cases. For example, in enzyme mechanisms the ionization states of a few key protein side chains are often critical. Suppose that two ionizable groups on the enzyme are critical for catalytic activity and that one of them needs to be protonated and the other deprotonated. The protolytic reactions can be written as

$$EH_2 \rightleftharpoons EH + H^+ \rightleftharpoons E + 2H^+$$

If only the species EH is catalytically active and the protolytic reactions are much more rapid than the other steps in the reaction, all of the rate constants that multiply the free enzyme concentration in the rate law have to be multiplied by the fraction of enzyme present as EH.

A. Calculate the fraction of free enzyme present as EH at a given pH. Your answer should contain the concentration of H^+ and the ionization constants of the two side chains, $K_{E1} = [E][H^+]/[EH]$ and $K_{E2} = [EH][H^+]/[EH_2]$.

B. Assume that ES in the Michaelis–Menten mechanism (Eq. 5-2) also exists in three protonation states, ESH_2, ESH, and ES, with only ESH being catalytically active. Calculate the fraction of the enzyme–substrate complex present as EHS. Designate the ionization constants as K_{ES1} and K_{ES2}.

C. Use the results of parts A and B to derive equations for the pH dependence of V_m, K_M, and V_m/K_M. Measurement of the pH dependence of the steady-state parameters permits determination of the ionization constants, and sometimes identification of the amino acid side chains.

CHAPTER 6

Ligand Binding to Macromolecules

6.1 INTRODUCTION

The binding of ligands to macromolecules is a key element in virtually all biological processes. *Ligands* can be small molecules, such as metabolites, or large molecules, such as proteins and nucleic acids. Ligands bind to a variety of *receptors*, such as enzymes, antibodies, DNA, and membrane-bound proteins. For example, the binding of substrates to enzymes initiates the catalytic reaction. The binding of hormones, such as insulin, to receptors regulates metabolic events, and the binding of repressors and activators to DNA regulates gene transcription. The uptake and release of oxygen by hemoglobin is essential for life. Indeed, compilation of a comprehensive list of biological processes in which ligand binding plays a key role would be a formidable task.

In this chapter, we shall discuss how to analyze ligand binding to macromolecules quantitatively for both simple and complex systems. We also will consider experimental methods that are used to study ligand binding. The application and importance of this analysis for biology will be illustrated through specific examples. The treatment presented will be adequate for most situations: more complete (and more complex) discussions of this topic are available (1–3).

6.2 BINDING OF SMALL MOLECULES TO MULTIPLE IDENTICAL BINDING SITES

The binding of a small molecule to identical sites on a macromolecule is a common occurrence. For example, enzymes frequently have several binding sites for substrates on a single molecule. Proteins have multiple binding sites for protons that are often essentially identical, for example, carboxyl or amino groups. Let us assume the simplest case, namely, a single ligand, L, binding to a single site on a protein, P:

$$L + P \rightleftharpoons PL \tag{6-1}$$

This equilibrium can be characterized by the equilibrium constant, K:

$$K = [PL]/([P][L]) \tag{6-2}$$

Binding equilibria involving macromolecules are conveniently characterized by a binding isotherm, the moles of ligand bound per mole of protein, r. For the above case,

$$r = \frac{[PL]}{[P] + [PL]} = \frac{K[L]}{1 + K[L]} \tag{6-3}$$

A plot of r versus [L] is shown in Figure 6-1a: It is a hyperbolic curve with a limiting value of 1 at high ligand concentrations, and when $r = 0.5$, $[L] = 1/K$.

If there are n identical binding sites on the protein, the binding isotherm for the macromolecule is simply the sum of those for each of the sites:

$$r = \frac{nK[L]}{1 + K[L]} \tag{6-4}$$

A plot of r versus [L] has the same shape as before, but now the limiting value of r at high ligand concentrations is n, the number of identical binding sites on the macro-molecule, and when $r = n/2$, $[L] = 1/K$. Thus, a study of ligand binding to a protein containing n identical binding sites permits determination of the number of binding sites and the equilibrium binding constant. In practice, alternative plots of the data are frequently used that yield a straight line plot and therefore can be analyzed more easily. With the availability of nonlinear least squares programs on desktop computers, this is not really necessary. However, it should be kept in mind that statistical analyses always fit the data—this does not necessarily mean that the fit is a good one. The quality of the fit must carefully be examined to be sure that the equations used and the fitting procedures are appropriate. In addition, it is extremely important that a very wide range of ligand concentration is used.

The most obvious recasting of Eq. 6-4 is to take its reciprocal:

$$1/r = 1/n + 1/(nK[L]) \tag{6-5}$$

A plot of $1/r$ versus $1/[L]$ is a straight line with an intercept on the [L] axis of $1/n$ and a slope of $1/nK$. Another possibility is to multiply Eq. 6-5 by [L]:

$$[L]/r = [L]/n + 1/(nK) \tag{6-6}$$

If $[L]/r$ is plotted versus [L], a straight line is obtained with a slope of $1/n$ and an intercept of $1/nK$. A common alternative plot is a Scatchard plot (4), named after George Scatchard, a pioneer in the study of small molecule binding to proteins. Rearrangement of Eq. 6-4 gives

$$r/[L] = nK - rK \tag{6-7}$$

A plot of $r/[L]$ versus r is a straight line, with the intercept on the x axis ($r/[L] = 0$) being equal to the number of binding sites, n, and the intercept on the y axis being equal to nK. Examples of these straight line plots are included in Figure 6-1.

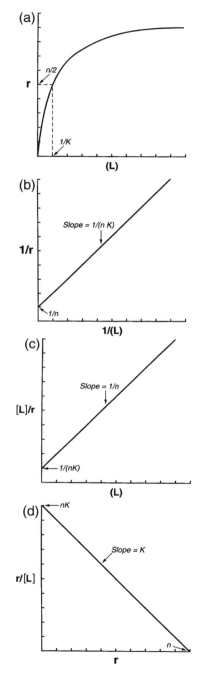

FIGURE 6-1. Plots of binding isotherms for n identical sites on a macromolecule according to (a) Eq. 6-4, (b) Eq. 6-5, (c) Eq. 6-6, and (d) Eq. 6-7. When $n = 1$, the curve in (a) obeys Eq. 6-3.

6.3 MACROSCOPIC AND MICROSCOPIC EQUILIBRIUM CONSTANTS

Before we proceed further with the analysis of ligand binding to macromolecules, it is important to understand the distinction between macroscopic and microscopic equilibrium constants. Consider, for example, a dibasic acid such as the amino acid glycine. Four possible protonation states are possible:

$$GH_2^+ = {}^+H_3NCH_2COOH$$

$$GH = H_2NCH_2COOH$$

$$GH' = {}^+H_3NCH_2COO^-$$

$$GH^- = H_2NCH_2COO^-$$

If a pH titration is carried out, the states GH and GH′ cannot be distinguished as they contain the same number of protons. The pH titration can be used to determine the *macroscopic* ionization constants:

$$K_1 = ([GH] + [GH'])[H^+]/[GH_2^+] \tag{6-8}$$

$$K_2 = [G^-][H^+]/([GH] + [GH']) \tag{6-9}$$

The two pK values determined by pH titration are p$K_1 = 2.35$ and p$K_2 = 9.78$ (25°C and zero salt concentration.)

If we consider the *microscopic* states of glycine, four *microscopic ionization constants*, k_i, are needed to characterize the system:

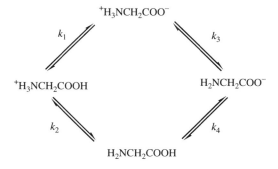

Note that these four microscopic ionization constants are not independent because this is a closed cycle. If the four microscopic ionization constants are written in terms of the concentrations, it can be seen that

$$k_1 k_3 = k_2 k_4 \tag{6-10}$$

This is an example of the principle of detailed balance. Whenever a closed cycle of reactions occurs, a relationship between the individual equilibrium constants such as Eq. 6-10 exists. The relationship between the macroscopic and microscopic ionization constants can easily be seen by reference to Eqs. 6-8 and 6-9, namely,

$$K_1 = k_1 + k_2 \tag{6-11}$$

and

$$K_2 = k_3 k_4/(k_3 + k_4) \tag{6-12}$$

Since the two macroscopic ionization constants and Eq. 6-10 provide only three relationships between the microscopic ionization constants, it is apparent that the microscopic ionization constants cannot be determined from a pH titration. In order to calculate the microscopic ionization constants, additional information is needed. In this case, it might be assumed that k_2 is equal to the ionization constant determined by pH titration of the methyl ester of glycine, namely, $pk_2 = 7.70$. This assumes that the ionization constant for protonation of the amino group is the same when either a proton or a methyl group is bound to the carboxyl, which is not unreasonable. With this assumption, the other three microscopic pK values can be calculated: $pk_1 = 2.35$, $pk_3 = 9.78$, and $pk_4 = 4.43$. These results indicate that the bottom state in the ionization scheme above is present at very low concentrations throughout a pH titration.

This is a simple illustration of the distinction between macroscopic and microscopic equilibrium constants. Experimental measurements usually only give information about macroscopic binding constants although exceptions exist. For example, the state of protonation of the nitrogen can be monitored by nuclear magnetic resonance (NMR) so that, in principle, a titration curve can be obtained for the amino group independently of that for the dibasic acid.

6.4   STATISTICAL EFFECTS IN LIGAND BINDING TO MACROMOLECULES

Let us return again to the binding of a ligand to multiple identical binding sites on a macromolecule and consider the matter of macroscopic and microscopic equilibrium constants. As a simple example, consider a macromolecule, P, with two identical binding sites; symbolically one of the binding sites will be on the left of P and the other on the right of P. The macroscopic equilibrium constants are

$$K_1 = ([PL] + [LP])/[L][P] \tag{6-13}$$

and

$$K_2 = [PL_2]/\{[L]([PL] + [LP])\} \tag{6-14}$$

In relating this discussion to the previous consideration of glycine, it should be noted that the glycine protolytic equilibria are characterized by equilibrium *dissociation* constants whereas equilibrium *association* constants are used here.

The equilibrium binding isotherm can be written as

$$r = \frac{[PL] + [LP] + 2[PL_2]}{[P] + [PL] + [LP] + [PL_2]} \tag{6-15}$$

If both the numerator and denominator are divided by [P], it can easily be seen that

$$r = \frac{K_1[L] + 2K_1K_2[L]^2}{1 + K_1[L] + K_1K_2[L]^2} \tag{6-16}$$

This does not look the same as Eq. 6-4. This is because the equilibrium constant in Eq. 6-4 is the microscopic equilibrium constant. In this case it is easy to see that $K_1 = 2K$ and $K_2 = K/2$. If these relationships are put into Eq. 6-16,

$$r = \frac{2K[L](1 + K[L])}{(1 + K[L])^2}$$

$$r = \frac{2K[L]}{1 + K[L]}$$

Thus, in the case of multiple identical binding sites, the binding isotherm is the same whether macroscopic or microscopic equilibrium constants are considered. The relationship between microscopic and macroscopic constants for this particular case is particularly simple. When both sites are empty, there are two possible sites available for the ligand so that the macroscopic equilibrium constant is twice as large as the microscopic constant. When both sites are occupied by ligand, there are two sites from which the ligand can dissociate; therefore, the macroscopic equilibrium constant is one-half of the microscopic constant.

If more than two identical binding sites are present, the relationship between the macroscopic and microscopic equilibrium constants can be determined in a similar manner. If a macromolecule has n identical binding sites, the relationship between the macroscopic equilibrium constant for binding to the ith site and the microscopic equilibrium constant is

$$K_i = \frac{\text{Number of free sites on P before binding}}{\text{Number of occupied sites on P after binding}} K$$

$$K_i = [(n - i + 1)/i] K \qquad i \geq 1 \tag{6-17}$$

We now return to consideration of a macromolecule with n identical binding sites.

The macroscopic equilibria are

$$L + P \rightleftharpoons PL$$

$$L + PL \rightleftharpoons PL_2 \tag{6-18}$$

$$\begin{array}{cc} \cdot & \cdot \\ \cdot & \cdot \\ \cdot & \cdot \end{array}$$

$$L + PL_{n-1} \rightleftharpoons PL_n$$

The corresponding macroscopic equilibrium constants are

$$K_1 = [PL]/([P][L]) \tag{6-19}$$

$$\begin{array}{cc} \cdot & \cdot \\ \cdot & \cdot \\ \cdot & \cdot \end{array}$$

$$K_n = [PL_n]/([L][PL_{n-1}])$$

The binding isotherm can be written as

$$r = \frac{[PL] + 2[PL_2] + \cdots + n[PL_n]}{[P] + [PL] + [PL_2] + \cdots + [PL_n]} \tag{6-20}$$

If we divide the numerator and denominator by [P] and use the definitions of the macroscopic dissociation constants for the ratios $[PL_i]/[P]$, Eq. 6-20 becomes

$$r = \frac{K_1[L] + 2K_1K_2[L]^2 + \cdots + nK_1K_2 \cdots K_n[L]^n}{1 + K_1[L] + K_1K_2[L]^2 + \cdots + K_1K_2 \cdots K_n[L]^n} \tag{6-21}$$

This equation describes the binding of a ligand to *n different* binding sites as no relationship has been assumed between the binding constants. If the sites are identical, the macroscopic constants are related to the microscopic or intrinsic binding constant by Eq. 6-17. If this relationship is inserted into Eq. 6-21, it becomes

$$r = \frac{nK[L] + \dfrac{2n(n-1)}{2!} K^2[L]^2 + \cdots + nK_n[L]^n}{1 + nK[L] + \dfrac{n(n-1)}{2!} K^2[L]^2 + \cdots + K_n[L]^n} \tag{6-22}$$

$$r = \frac{nK[L](1 + K[L])^{n-1}}{(1 + K[L])^n} = \frac{nK[L]}{1 + K[L]}$$

The binomial theorem has been used to obtain the final result; that is, the series of terms in the denominator that contain successive powers of [L] is recognized as the expansion of $(1 + K[L])^n$. Similarly, the expansion of $(1 + K[L])^{n-1}$ can be factored out of the numerator. This result is exactly the same as Eq. 6-4, which was obtained in a more intuitive manner.

To conclude this section, we consider the situation where a macromolecule has more than one set of independent identical binding sites for a ligand. In this case, the equilibrium binding isotherm is simply a sum of the terms given in Eq. 6-4, with the number of terms in sum being equal to the number of sets of binding sites:

$$r = \sum \frac{n_i K_i [L]}{1 + K_i [L]} \tag{6-23}$$

Here the n_i and K_i are the number of sites in each set and the intrinsic equilibrium constant for each set, respectively. Unless independent information is available about the structure of the macromolecule, it is usually preferable to use Eq. 6-21 to fit the data since no assumptions are made about the nature or number of the binding sites in this case.

Virtually any ligand binding isotherm can be fit to Eq. 6-21, but this is not a meaningful exercise unless the result can be related to the structure and/or the function of

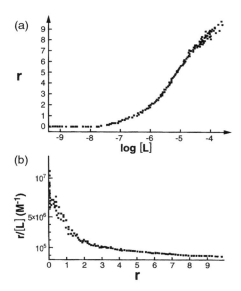

FIGURE 6-2. (a) Plot of r versus log [L] for the binding of laurate ion by human serum albumin. (b) The same data plotted as $r/[L]$ versus r (Scatchard plot). Adapted from I. M. Klotz, *Ligand–Receptor Complexes*, John Wiley & Sons, Inc. New York, 1997. Copyright © 1997 John Wiley & Sons, Inc. Reprinted by permission of John Wiley & Sons, Inc.

the macromolecule. For example, the binding isotherm for the binding of laurate ion to human serum albumin is shown in Figure 6-2 (5,6). The concentration range covered is so large that the concentration axis is logarithmic. The corresponding Scatchard plot also is shown in Figure 6-2. It is clear that saturation of the binding sites on serum albumin is never reached. The data in the figures, nevertheless, can be well fit with the assumption that $n = 10$. Clearly, this does not establish that 10 binding sites exist, nor regrettably does it provide information about the structure of the macromolecule. We will not dwell on this matter here. We will, however, return to the topic of multiple binding sites when discussing the concept of cooperativity.

6.5   EXPERIMENTAL DETERMINATION OF LIGAND BINDING ISOTHERMS

We will now make a digression to discuss briefly some of the experimental aspects of ligand binding studies. Many different experimental methods exist for determining the binding of ligands to macromolecules. The experimental methods used fall into two classes: (1) direct determination of the unbound ligand concentration; and (2) measurement of a change in physical or biological property of the ligand or macromolecule when the ligand binds. Generally, stoichiometry and binding constants can be determined reliably only if concentrations of the unbound ligand, the bound ligand, and unbound macromolecule can be determined.

The most straightforward method is equilibrium dialysis. This method is pictured schematically in the diagram in Figure 6-3. With this method, a solution is separated into two parts by a semipermeable membrane, which will not permit the macromolecule to cross but will permit the ligand to pass freely. The macromolecule is put on one side of the membrane and the ligand on both sides. The system is then permitted to come to equilibrium. At equilibrium the concentration of unbound ligand is the same on both sides of the membrane. (Strictly speaking, their thermodynamic activities are equal, but we will not worry about the difference between activity and con-

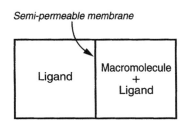

FIGURE 6-3. Schematic representation of an equilibrium dialysis experiment. The macromolecule is on one side of a semipermeable membrane and cannot pass through the membrane. The ligand can pass through the membrane and therefore is on both sides of the membrane. The concentration of the ligand on the side of the membrane that does not contain the macromolecule is equal to the concentration of unbound ligand.

centration here.) If the total amount of ligand is known, and the unbound amount of ligand is known, the amount of bound ligand can be determined from mass balance. The concentration of the macromolecule can be determined independently so that the binding isotherm can be calculated directly.

If equilibrium dialysis is to provide reliable data for analysis of the binding isotherm, it is usually necessary for the concentrations of bound and unbound ligand and macromolecule to be comparable. If essentially all of the ligand is bound, the amounts of unbound ligand and/or protein become very small and difficult to determine. On the other hand, if essentially no ligand is bound, calculation of the bound ligand becomes problematical. In the former case, the stoichiometry of binding can often be determined, but not the binding constant(s). In the latter case, it is virtually impossible to determine either the stoichiometry or binding constant(s). Another way of stating this is that the concentrations of all the species should be the same order of magnitude as the reciprocal of the binding constant so that experimental points can be determined over a wide range of r. This limitation becomes restrictive for very weak binding, primarily in terms of the large amount of material required. In the case of very tight binding, it may be difficult to work with the very dilute solutions required.

Equilibrium dialysis is an example of the more general method of determining the distribution between phases. This could, for example, be the equilibration between two nonmiscible solvents that partition the reactants, although this is rarely used. Various gel exclusion media such as Agarose are, however, often used. Such gels will exclude macromolecules from their beads, with the size of the macromolecule excluded depending on the property of the specific gel. The gel will include the ligand so that the gel bead is serving as semipermeable membrane, with equilibration occurring between the inside and outside of the bead. If a macromolecule is passed through a column equilibrated with a given ligand concentration, the unbound ligand concentration is equal to this concentration. The amount of ligand bound and the macromolecule concentration can be determined by analysis of the column effluent. Many variations on this type of experimental protocol exist.

In some cases, the concentration of unbound ligand is determined directly in the absence of a semipermeable membrane, for example, by a biological assay. In this case, the implicit assumption is that the rate of equilibration of the ligand and macromolecule occurs more slowly than the time for analysis. This is not necessarily the case, so such methods must be used with extreme caution.

Perturbations in the physical properties of the ligand are often used to determine binding isotherms. However, some care and caution must be exercised in doing so. The most frequent situation is when the optical absorption properties of the ligand are different for the bound and unbound ligand, or for the macromolecule with and without ligand. The technique of difference spectroscopy is often used. With this method, the difference in the optical absorption of a solution containing ligand and a solution containing the same concentration of ligand plus the macromolecule is determined. If the concentration of total ligand in the first solutions is $[L_T]$ and the concentration of free (unbound) ligand and bound ligand are $[L_f]$ and $[L_b]$, respectively, this difference, Δa, can be written as

$$\Delta a = \varepsilon_f[L_T] - \varepsilon_f[L_f] - \varepsilon_b[L_b] \tag{6-24}$$

$$\Delta a = (\varepsilon_f - \varepsilon_b)[L_b]$$

where the ε_i are extinction coefficients and the relationship $[L_T] = [L_f] + [L_b]$ has been used. This derivation assumes that the extinction coefficient of bound ligand is the same at all binding sites, which may or may not be valid. Consequently, this method is not as direct as the partitioning methods previously described. The difference extinction coefficient, $\varepsilon_f - \varepsilon_b$, can be determined by extrapolation to high ligand concentrations where the macromolecule is saturated with ligand. The difference absorbance measurements will then permit the determination of the bound ligand concentration, and the concentrations of the other species can be determined by mass balance. This method also works if the spectral change occurs in the macromolecule rather than in the ligand. The only difference is that the comparison solution contains macromolecule rather than protein, and the concentrations and extinction coefficients of the protein appear in Eq. 6-24. Perhaps not so obvious is the fact that this method also works if spectral changes occur in both the ligand and macromolecule when the ligand binds. We will not deal with this more complex situation, but the derivation is similar to the simpler cases discussed.

As a specific example of a difference spectrum titration, the difference absorbance is shown as a function of the ligand concentration for the binding of 2'-cytidine monophosphate to ribonuclease A in Figure 6-4 (7). Other optical properties such as fluo-

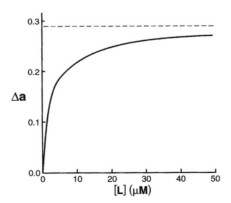

FIGURE 6-4. Plot of the absorbance difference at 288 nm, Δa, accompanying the binding of 2'-CMP to ribonuclease A versus the concentration of unbound 2'-CMP at pH 5.5, 25°C. The total enzyme concentration is 10^{-4} M. The difference extinction coefficient is 2.88×10^3 cm^{-1} M^{-1} and the binding constant is 2.96×10^5 M^{-1}. The curve was calculated from the data in D. G. Anderson, G. G. Hammes, and Frederick G. Walz, Jr., *Biochemistry* **7**, 1637 (1968). The dashed line is the difference absorbance when all of the enzyme has been converted to enzyme-2'-CMP.

rescence and circular dichroism can be used in a similar manner. Finally, it is sometimes possible to monitor a specific site on the macromolecule with physical methods such as NMR or other types of spectroscopy. If changes occur when ligand binds, the concentrations of empty and occupied sites can be determined and translated into binding isotherms.

The determination of high-quality binding isotherms is not trivial, and this discussion does not do full justice to the topic. One of the difficulties not discussed that should be mentioned is the occurrence of nonspecific binding. It is commonplace for nonspecific binding to occur along with binding of the ligand to the site(s) of interest. Macromolecules have many different side chains that can attract ligands, for example, by charge–charge interactions. This nonspecific binding is characterized by binding isotherms that do not level off at a specific value of n due to a large number of such sites characterized by very weak (small) binding constants. More complete treatises concerned with the experimental aspect of the binding of ligands to macromolecules are available (cf. Ref. 3).

6.6 BINDING OF CRO REPRESSOR PROTEIN TO DNA

As an example of a ligand binding study, we will consider some of the results obtained in the studies of the binding of the Cro repressor protein to DNA. DNA transcription is regulated by the binding of proteins to DNA. The Cro repressor is a protein that plays a regulatory role in λ phage. The binding of this protein to DNA has been examined by many different methods (8–10), and the three-dimensional structure of a Cro protein–DNA complex has been determined (11). The Cro protein binds as a dimer, and amino acids form DNA sequence specific hydrogen bonds with exposed parts of DNA bases. Both the thermodynamics and kinetics of Cro protein binding to DNA have been studied extensively.

The binding of Cro protein to DNA is difficult to study because the binding is very tight. In order to study the binding, a very sensitive filter binding assay was used. Under certain conditions, linear duplex DNA passes through nitrocellulose filters, while a protein or protein–DNA complex is retained. Radioactive (^{32}P) DNA was prepared and mixed with Cro protein at very low concentrations (as low as < 1ng/mL). The reaction mixture was filtered within a few seconds, and from the radioactivity bound to the filter, the concentration of Cro protein–DNA could be calculated. Since the total concentrations of Cro protein and DNA are known, the equilibrium concentrations of all species can be calculated. The binding of Cro protein to various operator regions of λ phage DNA was studied. The operator is 17 base pairs long, but the DNAs investigated contained an extra 2 base pairs at each end. In addition, a consensus operator was constructed that had the consensus sequence of the six λ operator regions examined. The consensus sequence is identical to the OR3 operator, except for a few base pairs on the interior of the sequence.

The equilibrium association constants obtained are given in Table 6-1. The stoichiometry of the reaction was determined to be one to one. The association constants are very large, with the largest being for the consensus and OR3 operator. The OR3 operator was previously determined to be the preferred binding site for Cro protein. The binding of Cro operator to nonspecific DNA was also determined and found to be a linear function of the length of the DNA with a binding constant of 6.8×10^5 M^{-1} per base pair, considerably weaker than the specific binding, as expected. The rate constants characterizing the reaction between the DNA operator sequences and Cro protein were also determined. The dissociation rate constants are included in Table 6-1. The association rate constants that could be measured were all approximately 3×10^8 M^{-1} s^{-1}, the value expected for a diffusion controlled reaction. Every time a Cro protein and the operator DNA collide, a complex is formed.

The binding of Cro protein to long DNAs containing the operator region was also studied. The length of DNA varied from 73 to 2410 base pairs. For all except the shortest DNA, the equilibrium constant was about the same, 6.7×10^{10} M^{-1}. However, the association and dissociation rate constants increase as the length of DNA increased, leveling off at values of about 4.5×10^9 M^{-1} s^{-1} and 1.7×10^{-2} s^{-1}, respectively. This very high second order rate constant suggests that every time Cro protein collides with DNA, it binds very tightly. This result is puzzling as it would be expected that Cro protein would have to sample various parts of the DNA until it found the operator and bound tightly. An explanation for these results is that Cro protein binds to the DNA on every collision and then diffuses rapidly along the DNA chain until it encounters the operator region. The rapid sliding of DNA binding protein along the DNA chain until it finds the correct place for a specific interaction appears to occur in other systems, but other mechanisms may also be operative.

Calorimetry was also carried out on the binding reaction (10). The results obtained indicate that the association of Cro protein with nonspecific DNA at 15°C is characterized by $\Delta H° = 4.4$ kcal/mol, $\Delta S° = 49$ cal/(mol·K), $\Delta G° = -9.7$ kcal/mol, and $\Delta C_P \approx 0$. The parameters obtained with the OR3 DNA are quite different, $\Delta H° = 0.8$ kcal/mol, $\Delta S° = 59$ cal/(mol·K), $\Delta G° = -16.1$ kcal/mol, and $\Delta C_P = -360$ cal/(mol·K). In both cases, the favorable free energy change is entropy driven.

The specific molecular interactions between the DNA and Cro protein have been probed by studying the binding of Cro protein to OR1 DNA, 21 base pairs as above, with systematic base substitutions along the entire DNA chain (9). Experiments were

TABLE 6-1. Binding Constants and Dissociation Rate Constants at 273 K for Cro Protein-λ DNA

Operator	$K(M^{-1})$	k_d (s^{-1})
Consensus	8.3×10^{11}	7.7×10^{-5}
OR3	5.0×10^{11}	1.7×10^{-4}
OR2	8.3×10^9	1.2×10^{-2}
OR1	1.2×10^{11}	1.7×10^{-3}
OL1	6.7×10^{10}	2.2×10^{-3}
OL2	3.7×10^{10}	2.6×10^{-3}
OL3	1.9×10^{10}	9.2×10^{-3}

Source: J. G. Kim, Y. Takeda, B. W. Matthews, and W. F. Anderson, *J. Mol. Biol.* **196**, 149 (1987).

also done with point mutations in the protein of amino acids thought to be involved in the binding interaction. The changes in free energy were then interpreted as due to specific interactions between the protein and DNA, and a structural map of the interactions was proposed. A high-resolution crystal structure of a Cro protein–DNA complex determined some years later provides a definitive description of the molecular interactions (11). Many features proposed on the basis of the binding/mutation studies proved to be correct although some were not. As often stated in this text, molecular interpretations of thermodynamic studies must be viewed with caution. However, in retrospect, all of the mutation studies can be understood in terms of the crystal structure. A view of the overall complex is presented in Color Plate VI. Each Cro protein monomer consists of three α-helices and three β-strands. Only the helix-turn-helix portion of the protein makes direct contact with the DNA bases: the α_3-helix is inserted into the major groove of DNA, and its interactions with DNA bases account for the tight operator binding. A schematic map of the DNA base–Cro protein interactions is shown in Figure 6-5. Multiple hydrogen bonds are formed, although other types of interactions are also important. Both the protein and DNA conformations are altered when the complex is formed. Many other interesting details of the mo-

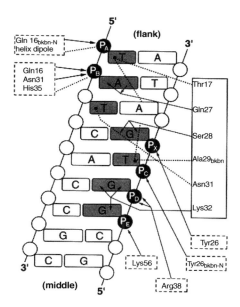

FIGURE 6-5. Schematic diagram of the interactions between the Cro protein and λ operator DNA. The interactions for one of the polypeptide chains are shown as the interactions with the other polypeptide chain of the dimeric Cro protein are symmetric. Hydrogen bonds are shown as continuous lines with arrows pointing from the donor to the acceptor. Broken lines are van der Waals contacts. bkbn indicates a contact with the protein backbone, and the dotted lines are presumed electrostatic interactions. Reproduced from R. A. Albright and B. W. Matthews, *J. Mol. Biol.* **280**, 137 (1998). Reprinted with permission from Academic Press, Inc.

lecular interactions can be inferred from the crystal structure but will not be considered here.

This study is an elegant demonstration of the type of information that can be obtained from thermodynamic, kinetic, and structural studies of ligand binding.

6.7 COOPERATIVITY IN LIGAND BINDING

We now return to the subject of macromolecules with multiple ligand binding sites. One of the most famous and best studied examples is hemoglobin. In Figure 6-6, the binding isotherm is shown for the binding of oxygen to myoglobin and hemoglobin. Hemoglobin contains four identical binding sites for oxygen. Myoglobin contains a single binding site for oxygen. The concentration of unbound oxygen is expressed in pressure units of mmHg and rather than r, the percent saturation (r/n) is shown so that the binding isotherms can be compared directly even though $n = 1$ for myoglobin and 4 for hemoglobin. The binding isotherm for myoglobin can be fit to Eq. 6-4 with $K = 0.23$ $(mmHg)^{-1}$ {$K = [MyO_2]/([My][O_2])$, where My is myoglobin}. The data for hemoglobin clearly cannot be fit so simply, as the binding isotherm is not hyperbolic. Equation 6-21 can be used, however. Before doing so it is useful to take into account the statistical effects associated with identical binding sites, that is, express Eq. 6-21 in terms of intrinsic binding constants rather than macroscopic binding constants by use of Eq. 6-17. With this substitution, Eq. 6-21 becomes

$$r = \frac{4K_1[O_2] + 12K_1K_2[O_2]^2 + 12K_1K_2K_3[O_2]^3 + 4K_1K_2K_3K_4[O_2]^4}{1 + 4K_1[O_2] + 6K_1K_2[O_2]^2 + 4K_1K_2K_3[O_2]^3 + K_1K_2K_3K_4[O_2]^4} \quad (6\text{-}25)$$

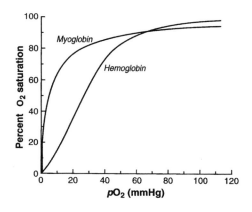

FIGURE 6-6. Binding of oxygen by myoglobin and hemoglobin at pH 7 and 38°C. The pressure of O_2 is a measure of the concentration of unbound O_2. Adapted from F. Daniels and R. A. Alberty, *Physical Chemistry*, 4th edition, John Wiley & Sons, Inc., New York, 1975. Copyright © 1975 John Wiley and Sons, Inc. Reprinted by permission of John Wiley & Sons, Inc.

If this equation is used to fit the data obtained at pH 7.4, 25°C, 0.1 M NaCl, the binding constants obtained are 0.024, 0.077, 0.083, and 7.1 $(mmHg)^{-1}$ (12). Note that the intrinsic binding constant becomes larger as each oxygen is bound. This is an example of cooperative binding. When an oxygen binds to hemoglobin, it increases the affinity of hemoglobin for the next oxygen. The result is a sigmoidal binding isotherm rather than a hyperbolic isotherm. This has important physiological consequences as it permits oxygen to be picked up and released over a very narrow range of oxygen pressure. Cooperative binding such as this is found frequently because it permits the biological activity to be regulated over a very narrow range of concentration.

When the binding of oxygen by hemoglobin was first studied, the binding isotherm was fit by the empirical equation

$$r/n = \frac{[L]^{\alpha}/K^{\alpha}}{1 + [L]^{\alpha}/K^{\alpha}} \qquad (6\text{-}26)$$

Where the equilibrium constant is expressed as a dissociation constant and α is an empirical parameter called the Hill coefficient obtained from experiment by plotting $\ln[(r/n)/(1 - r/n)]$ versus $\ln[L]$ (13). {Note that $(r/n)/(1 - r/n) = [L]^{\alpha}/K^{\alpha}$.} As might be expected, such plots are linear over a limited range of ligand concentration. In the case of hemoglobin, α is about 2.5 and depends on the specific experimental conditions. The use of this equation does not have a physical meaning. It assumes that the binding equilibrium is

$$\text{Hemoglobin} + \alpha O_2 \rightleftharpoons \text{Hemoglobin}(O_2)_{\alpha}$$

Obviously the number of binding sites on hemoglobin must be an integer. Nevertheless, cooperative binding is frequently characterized by a Hill coefficient. The closer the Hill coefficient is to the actual number of binding sites, the more cooperative the binding.

It is also possible to have binding isotherms in which the binding constants decrease as successive ligands bind. This is usually termed *negative cooperativity*, or *anticooperativity*, both oxymorons of a sort. It should be remembered that if macroscopic equilibrium binding constants are used, the binding constants decrease as each ligand is added to the macromolecule even if the sites are equivalent. This is the statistical effect embodied in Eq. 6-17. Anti- or negative cooperativity therefore means the binding constants decrease more than the statistical effect expected for equivalent sites. An example of negative cooperativity is shown in Figure 6-7 for the binding of cytidine 3′-triphosphate (CTP) to the enzyme aspartate transcarbamoylase (14). This enzyme catalyzes the carbamoylation of aspartic acid at a branch point in metabolism that eventually leads to the synthesis of pyrimidines. When the concentration of CTP becomes high, it inhibits aspartate transcarbamoylase and shuts down the metabolic pathway for its synthesis. As can be seen from the Scatchard plot presented, the data

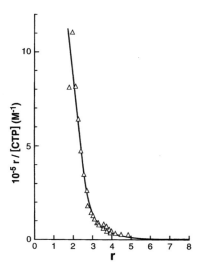

FIGURE 6-7. Scatchard plot for the binding of CTP to aspartate transcarbamoylase at 4°C and pH 7.3 in the presence of 2 mM carbamoyl phosphate and 10 mM succinate. Adapted from S. Matsumoto and G. G. Hammes, *Biochemistry* **12**, 1388 (1973). Copyright © 1973 American Chemical Society.

clearly do not conform to the straight line expected for independent equivalent binding sites. The total number of binding sites appears to be six, although extrapolation to this number is not precise. In this case, the data were fit to a model of two sets of three independent binding sites. The intrinsic binding constants obtained were 7.1×10^5 M^{-1} and 4.4×10^3 M^{-1}. The data could be fit equally well to Eq. 6-21, but the data were not sufficient to define all six constants well. Aspartate transcarbamoylase is known to contain six identical binding sites for CTP, which exist as three dimers of identical polypeptide chains. When CTP binds to one of the sites on a dimer, it weakens the binding for the second CTP. Interestingly, the binding of aspartate to this enzyme exhibits positive cooperativity, and the extent of the cooperativity is modulated by the binding of CTP, as will be discussed later.

The presence of positive or negative cooperativity can readily be diagnosed from the binding isotherm. The shapes of the various plots commonly used are shown in Figure 6-8 for no cooperativity, positive cooperativity, and negative cooperativity. The type of cooperativity occurring can readily be diagnosed from these plots. It is particularly obvious in Scatchard plots as no cooperativity gives a straight line, a maximum in the plot is observed for positive cooperativity, and a concave curve is found for negative cooperativity. For some studies, log[L] is used in order that a wide range of concentrations can be represented in a single plot (Fig. 6-8b).

When positive and negative cooperativity are observed, one must determine if this is due to a preexisting difference between the binding sites, or if the binding of one

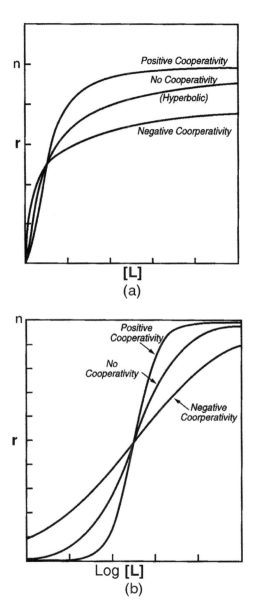

FIGURE 6-8. Schematic representations of equilibrium binding data demonstrating no cooperativity, positive cooperativity, and negative cooperativity. In these figures, r is the moles of ligand bound per mole of protein and [L] is the concentration of unbound ligand.

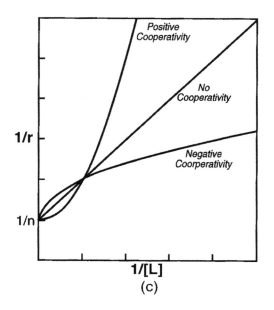

(c)

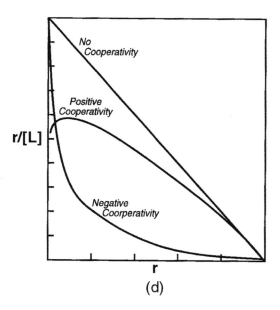

(d)

FIGURE 6-8. (*Continued*)

ligand alters the binding affinity of the remaining sites for the ligand. Usually, but not always, the latter is true, as for the cases cited above. The possibility also exists that both positive and negative cooperativity could occur in a single binding isotherm.

6.8 MODELS FOR COOPERATIVITY

Thus far we have been concerned with fitting data to binding isotherms and developing criteria with regard to whether the binding is cooperative. However, the ultimate goal is to relate the experimental data to molecular structure. This requires the development of theoretical models that relate structure to ligand binding isotherms. Two limiting models have been developed to explain cooperative ligand binding to proteins. Both models are based on the general hypothesis that cooperativity is the result of alterations in the interactions between polypeptide chains through conformational changes in the macromolecule. One of the models assumes a concerted or global conformational change, whereas the other assumes a sequential change in the conformation of each polypeptide chain or chains that contain a ligand binding site.

We first consider the concerted model that has been developed by Monod, Wyman, and Changeux (15). This model (MWC) is based on three assumptions: (1) The protein consists of two or more identical subunits, each containing a site for the ligand; (2) at least two conformational states, usually designated as R and T states, are in equilibrium and differ in their affinities for the ligand; and (3) the conformational changes

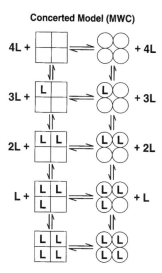

FIGURE 6-9. Schematic representation of the Monod–Wyman–Changeux model for a four subunit protein. The squares and circles designate different subunit conformations of the protein, and L is the ligand.

of all subunits occur in a concerted manner (conservation of structural symmetry). A schematic illustration of the MWC model for a four subunit protein is shown in Figure 6-9. A sigmoidal binding isotherm can be generated from this model in the following way. In the absence of ligand, the protein exists largely in the T state (the square conformation), but substrate binds preferentially to the R state (the circular conformation). When ligand binds, it shifts the protein from the T to the R state. Thus, at low ligand concentrations, the protein is primarily in the T state whereas at high ligand concentration, the protein is largely in the R state. This shift in equilibria can give rise to a sigmoidal binding isotherm, or positive cooperativity. This model cannot, however, explain anti (negative) cooperativity.

The quantitative development of this model is complex, although not difficult, and will only be given in outline form here. The scheme for ligand binding can be represented as

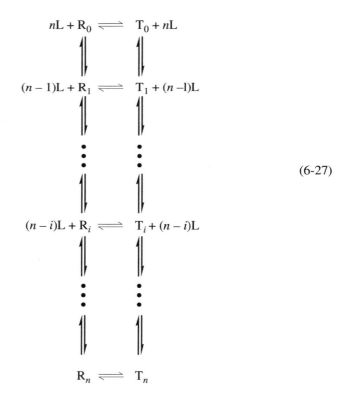

$$(6\text{-}27)$$

where R_0 and T_0 are the two different conformational states in the absence of ligand, and R_i and T_i designate their complexes with i molecules of L. Three constants are needed to specify the equilibrium binding isotherms: the intrinsic dissociation constants for ligand binding to the R and T states, and the equilibrium constant for the ratio of the R_0 to T_0 states. These constants can be written as

$$L_0 = [T_0]/[R_0]$$

$$K_R = [(n - i + 1)/i][R_{i-1}][L]/[R_i]$$

$$K_T = [(n - i + 1)/i][T_{i-1}][L]/[T_i] \qquad (6\text{-}28)$$

The fraction of sites occupied by the ligand, Y, can be expressed as

$$Y = \frac{r}{n} = \frac{([R_1] + 2[R_2] + \cdots + n[R_n]) + ([T_1] + 2[T_2] + \cdots + n[T_n])}{n\{([R_0] + [R_1] + \cdots + [R_n]) + ([T_0] + [T_1] + \cdots + [T_n])\}} \qquad (6\text{-}29)$$

$$Y = \frac{L_0 c\alpha(1 + c\alpha)^{n-1} + \alpha(1 + \alpha)^{n-1}}{L_0(1 + c\alpha)^n + (1 + \alpha)^n}$$

where $\alpha = [L]/K_R$ and $c = K_R/K_T$. The transformation of the first part of Eq. 6-29 to the second part requires the use of the binomial theorem and will not be detailed here. The nature of the binding isotherm depends on the values of L_0 and c. A hyperbolic isotherm is obtained when the ligand binds equally well to both conformations, $c = 1$, and when L_0 and c are both either very large or very small. However, when L_0 is large and c is small, sigmoidal binding isotherms occur as illustrated in Figure 6-10. When c is very small, Eq. 6-29 becomes

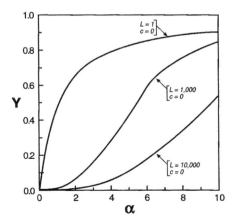

FIGURE 6-10. Plot of the fraction of sites bound by ligand, Y, versus α (= $[L]/K_R$) for various values of c and L according to Eq. 6-29. A sigmoidal binding isotherm is generated when L is large and c is small.

$$Y = \frac{\alpha(1 + \alpha)^{n-1}}{L_0 + (1 + \alpha)^n} \tag{6-30}$$

With this limiting case, it can easily be seen that a hyperbolic isotherm is found when L_0 is small, whereas a sigmoidal isotherm is predicted when L_0 is large.

The basic assumptions of an alternative model developed by Koshland, Nemethy, and Filmer (KNF) are the following (16): (1) Two conformational states, A and B, are available to each subunit; (2) only the subunit to which the ligand is bound changes its conformation; and (3) the ligand induced conformational change in one subunit alters its interactions with the neighboring subunits. The strength of the subunit interactions may be increased, decreased, or stay the same. The result of this change in subunit interactions is that the binding of the next ligand can be weaker, stronger, or the same as the binding of the previous ligand. Clearly, this model can produce either positive or negative cooperativity—or a hyperbolic binding isotherm. This sequential model is shown schematically in Figure 6-11 for a four subunit protein in a square configuration.

Calculation of the binding isotherm for the KNF model is complex and will not be presented here. The basic parameters are the intrinsic binding constant, an equilibrium constant characterizing the conformational change that occurs, and constants characterizing the subunit interactions, AB and BB. (The AA interaction is taken as the reference state so it does not appear in the calculation.) The binding isotherm that results is identical in form to Eq. 6-21, except that the appropriate statistical corrections are included so that the intrinsic binding constant appears. The effective binding constant multiplying each successive power of L may increase, decrease, or stay the same, depending on the nature of the subunit interactions, so that positive, negative, or mixed cooperativity is possible.

The MWC and KNF models are limiting cases of a more general scheme shown in Figure 6-12. This figure illustrates a general mechanism involving a tetrameric protein and only two conformational states for each subunit. The real situation is somewhat more complex, as the permutations of the ligand among the subunits for a given conformational state are not shown. The extreme right- and left-hand columns enclosed by dashed lines represent the MWC mechanism, whereas the diagonal, enclosed by

Sequential Model (KNF)

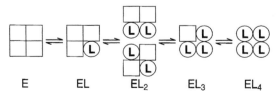

FIGURE 6-11. Schematic representation of the Koshland–Nemethy–Filmer model for a four subunit (square) protein. The squares and circles designate different subunit conformations, and L is the ligand. Note that two structures are shown for the intermediate with two ligands bound as the subunit interactions are different for these two species.

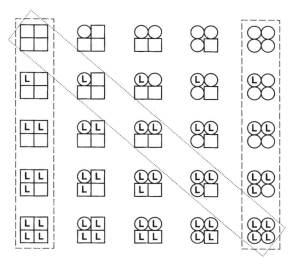

FIGURE 6-12. Schematic illustration of a general allosteric model for the binding of a ligand to a four subunit protein. The squares and circles designate different subunit conformations. The portion enclosed by dashed lines is the Monod–Wyman–Changeux model, whereas that enclosed by dotted lines is the Koshland–Nemethy–Filmer model. For the sake of simplicity, the permutations of the ligand among the subunits and the free ligand are omitted.

dotted lines, represents the KNF model. Thus, these two models are limiting cases of an even more complex scheme. As might be suspected, distinguishing among these models when positive cooperativity occurs is not a simple task.

6.9 KINETIC STUDIES OF COOPERATIVE BINDING

Thus far we have confined the discussion to ligand binding at equilibrium. The binding process and the cooperativity that may occur play an important role in many biological processes, as evidenced by hemoglobin, membrane receptor binding, and operator binding to DNA. The role of cooperativity in regulating such reactions has been well established. However, many enzyme reactions are also regulated through cooperative binding processes. In such cases, the rate of the enzymatic reaction is usually measured, often the initial velocity, and it is observed that a plot of the initial velocity versus the substrate concentration exhibits cooperativity. An example already discussed is aspartate transcarbamoylase: A plot of the initial velocity versus the aspartate concentration is sigmoidal so that small changes in the concentration of aspartate can cause significant changes in the rate of the enzymatic reaction.

For many enzymes, plots of the initial velocity versus the substrate concentration are essentially identical in form with the binding isotherm. This suggests that the binding steps prior to the rate determining step are rapid and reversible and that the turnover numbers (catalytic efficiency) for all of the binding sites are identical. If this is the situation, the initial velocity, v, can readily be related to the binding isotherm:

$$v = V_m \, r \tag{6-31}$$

where V_m is the maximum velocity of the enzyme reaction expressed as the product of the molar concentration of enzyme and the turnover number. This simple analysis seems to be sufficient for many cases, and if it is, the treatment of the rate data parallels what was done for equilibrium binding, including the methods of plotting and fitting the data. As might be expected both positive and negative cooperativity are observed.

The interpretation of kinetic data can, however, be more complex. The turnover numbers for different sites could be different. For example, in terms of the MWC model, the R and T forms might have different catalytic activities although usually the assumption is made that only the R form is enzymatically active. For the KNF model each of the ligand binding sites might have a different catalytic activity. In fact, these complications are rarely included, or justified, in the data analysis. Finally, it should be noted that apparent cooperativity in the rate of an enzymatic reaction can arise purely from special kinetic situations and may have nothing to do with cooperative binding of substrate. A few such situations have been well documented, but we will not consider such complications further. This is just a reminder that kinetic measurements are not a substitute for equilibrium binding studies. They may provide similar information in some cases, but they are inherently more difficult to interpret. Of course, conversely, kinetic studies can provide dynamic information that cannot be obtained from equilibrium measurements.

6.10 ALLOSTERISM

Thus far we have considered only cases where a single ligand binds to a macromolecule, and important biological control can occur through cooperative interactions for a single ligand. These are *homotropic* interactions. However, in many instances regulation occurs through the binding of a second ligand that influences the binding of the first ligand. These are called *heterotropic* interactions. Regulatory control by reaction of a second ligand is termed *allosterism*, and the second binding site is called an *allosteric* site. Two specific examples will be discussed to illustrate the principles involved: hemoglobin and aspartate transcarbamoylase.

The association of oxygen with hemoglobin is strongly pH dependent as shown in Figure 6-13. Note that the oxygenation isotherm becomes more sigmoidal as the pH is lowered. Thus, the binding of protons to hemoglobin clearly affects oxygen binding. In fact, the addition of protons decreases the amount of oxygen bound, and vice versa. This reciprocal relationship is called the Bohr effect. The effect of proton binding is typical allosteric regulation, in this case of oxygen binding. How is this accommodated in the models we have discussed previously? For the MWC model, allosteric effectors are assumed to bind selectively to the R or T conformation. An inhibitor, I, such as the proton in the case under consideration, would bind selectively to the T conformation. This shifts the equilibrium from the R to the T state, which effectively changes the equilibrium constant L_0 to

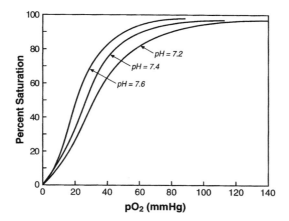

FIGURE 6-13. The effect of pH on the oxygenation of hemoglobin (the Bohr effect). The percent of saturation by oxygen is plotted versus the pressure of O_2. Adapted from R. E. Benesch and R. Benesch, *Adv. Prot. Chem.* **28**, 211 (1974). Reprinted with permission of Academic Press, Inc.

$$L_0' = L_0(1 + [I]/K_I)^n \qquad (6\text{-}32)$$

where K_I is the intrinsic dissociation constant for the binding of inhibitor, and it has been assumed that a single proton binds to each of the four subunits. Insertion of this relationship into Eq. 6-29 can quantitatively account for the Bohr effect. The KNF model can also explain the observations by assuming that binding of the inhibitor alters the subunit interactions and conformational changes, effectively altering the binding constants for oxygen as successive ligands are bound. Organic phosphates are also strong effectors of oxygen binding and bind preferentially to deoxyhemoglobin.

An activator, A, can enhance ligand binding by binding selectively to the R conformation. This effectively changes the equilibrium constant L_0 to

$$L_0' = \frac{L_0}{(1 + [A]/K_A)^n} \qquad (6\text{-}33)$$

where K_A is the intrinsic dissociation constant for the binding of activator.

Which of the models for allosterism best accommodates the known data for hemoglobin? The nature of the conformational change occurring when oxygen binds is known from structural studies of deoxy- and oxyhemoglobin. The hemoglobin studied contains four polypeptide chains, two α chains and two β chains. The α and β chains are similar but not identical. When oxygen binds, salt bridges are broken between the polypeptide chains, and all four chains rotate slightly to accommodate the movement of iron into the plane of the heme. The iron moves only a few tenths of an angstrom, and two of the polypeptides move about 7 Å closer (17). These structural changes and the binding data fit quite well to the MWC model. However, there is evidence that in addition to this major conformational change, sequential conformational changes occur in the subunits. Therefore, it is likely the case that both models are needed to accommodate the data, and multiple conformational changes occur. The structure of hemoglobin and the proposed major concerted conformational change are shown schematically in Color Plate VII.

As previously indicated, aspartate transcarbamoylase is a key enzyme in the pathway for pyrimidine biosynthesis. It is subject to inhibition by CTP and to activation by ATP. The allosteric binding of CTP makes the dependence of the rate on aspartate concentration more sigmoidal, whereas binding of the activator, ATP, makes the curve less sigmoidal (18). This is a prototype for feedback inhibition in metabolism and is shown schematically in Figure 6-14. An additional wrinkle in the regulatory process is that the binding of both CTP and ATP to the enzyme display negative cooperativity, and these two ligands compete for the same binding site. The effect of ATP and CTP on the binding of aspartate can readily be accommodated by the MWC model with the assumption that CTP binds selectively to the T conformation and ATP to the R conformation. This assumes that aspartate binds selectively to the R conformation. This model, however, cannot accommodate the negative cooperativity observed in the binding of CTP and ATP. The KNF model also can accommodate these results through alterations in the interactions between subunits.

Again, the structure of aspartate transcarbamoylase is known. It consists of two trimers of catalytic sites, with three dimers of regulatory sites at the interface of the trimers (19). A threefold rotational axis is present, and it has been shown that conversion of the putative R to T forms involves rotation around this axis and alteration of the interactions between the regulatory and catalytic subunits. The MWC model accommodates much of the data, but the data also require sequential conformational changes in the subunits. Thus, the conclusion is similar to that for hemoglobin. A major conformational change consistent with R and T conformations appears to occur,

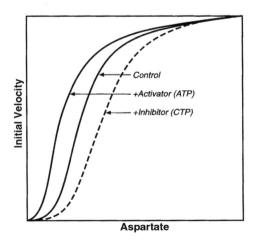

FIGURE 6-14. Schematic representation of the dependence of the initial velocity of the reaction catalyzed by aspartate transcarbamoylase on the concentration of aspartate, a substrate. The effect of an allosteric inhibitor, CTP, and of an allosteric activator, ATP, are shown.

but additional conformational changes more localized and sequential in nature also occur. The regulatory process, therefore, seems to require multiple conformations and the interplay between global and local conformational changes. The use of multiple conformational changes enhances the versatility and sensitivity of the regulatory process. The structure of aspartate transcarbamoylase and the nature of the concerted conformational change are shown in Color Plate VIII. The top of the figure is the T state with the catalytic trimers in blue and white at the top and bottom of the structure. Two of the regulatory dimers are at the sides of the structure in yellow. The third dimer is in the back of the structure. The bottom structure is the R state. The movement of the subunits with respect to each other can clearly be seen.

This concludes our discussion of ligand binding in biology. We have developed the theoretical and experimental framework and have provided several examples of applications to biological systems. This, and further reading of the literature, should permit the interested reader to develop specific applications as needed.

REFERENCES

1. G. G. Hammes, *Enzyme Catalysis and Regulation*, Academic Press, New York, 1982.

2. C. R. Cantor and P. R. Schimmel, *Biophysical Chemistry*, Parts I, II, and III, W. H. Freeman, San Francisco, 1980.

3. I. M. Klotz, *Ligand–Receptor Complexes*, John Wiley & Sons, New York, 1997.

4. G. Scatchard, *Ann. N.Y. Acad. Sci.* **51**, 660 (1949).

5. A. O. Pedersen, B. Hust, S. Andersen, F. Nielsen, and R. Brodersen, *Eur. J. Biochem.* **154**, 545 (1986).

6. R. Brodersen, B. Honoré, A. O. Pedersen, and I. M. Klotz, *Trends Pharm. Sci.* **9**, 252 (1988).

7. D. G. Anderson, G. G. Hammes, and Frederick G. Walz, Jr., *Biochemistry* **7**, 1637 (1968).

8. J. G. Kim, Y. Takeda, B. W. Matthews, and W. F. Anderson, *J. Mol. Biol.* **196**, 149 (1987).

9. Y. Takeda, A. Sarai, and V. M. Rivera, *Proc. Natl. Acad. Sci. USA* **86**, 439 (1989).

10. Y. Takeda, P. D. Ross, and C. P. Mudd, *Proc. Natl. Acad. Sci. USA* **89**, 8180 (1992).

11. R. A. Albright and B. W. Matthews, *J. Mol. Biol.* **280**, 137 (1998).

12. I. Tyuma, K. Imai, and K. Shimizu, *Biochemistry* **12**, 1491 (1973).

13. A. V. Hill, *J. Physiol. (London)* **40**, iv (1910).

14. S. Matsumoto and G. G. Hammes, *Biochemistry* **12**, 1388 (1973).

15. J. Monod, J. Wyman, and J.-P. Changeux, *J. Mol. Biol.* **12**, 88 (1965).

16. D. E. Koshland, Jr., G. Nemethy, and G. Filmer, *Biochemistry* **5**, 365 (1966).

17. J. M. Friedman, *Science* **228**, 1273 (1985).

18. J. C. Gerhart and A. B. Pardee, *J. Biol. Chem.* **237**, 819 (1962).

19. W. N. Lipscomb, *Adv. Enzymol.* **68**, 67 (1994).

PROBLEMS

6-1. The following data were obtained for the binding of ADP to an ATPase.

r	[ADP] (μM)
0.500	0.719
0.694	1.23
1.06	2.36
1.22	3.30
1.32	4.55
1.63	8.81
1.69	16.32
1.91	27.15
2.21	40.62
2.15	79.75
2.40	115.5

How many binding sites are present per mole of enzyme and what is the intrinsic binding constant? The concentration in the table is unbound ADP. (Nonspecific binding occurs, which is not uncommon. You will have to decide how to handle this problem.)

6-2. Some typical equilibrium dialysis data for the binding of a ligand to a macromolecule are given below. The total concentration of the macromolecule is 0.500 μM.

Total Ligand Concentration (μM)

Side Without Macromolecule	Side With Macromolecule
0.158	0.436
0.395	0.960
0.890	1.83
2.37	3.78
4.00	5.60
6.18	7.91
8.12	9.90

Determine the number of binding sites on the macromolecule and the intrinsic binding constant.

6-3. The following initial velocities, v, were measured for the carbamoylation of aspartic acid by carbamoyl phosphate as catalyzed by the enzyme aspartate transcarbamoylase. The concentration of carbamoyl phosphate was 1 mM, and the concentration of aspartate was varied.

v (arbitrary units)	Aspartate (mM)
0.90	2.0
1.60	3.0
2.25	4.0
3.20	5.0
3.65	6.0
4.70	8.0
5.05	10.0
5.25	12.0
5.80	15.0
6.00	17.0
6.05	20.0

A. Assume the initial velocity is related to r by Eq. 6-31 and construct a Hill plot of the data. The slope provides a lower bound to the number of aspartate binding sites.

B. In fact, six aspartate binding sites are present per mole of protein. Using this information and the data, make a table of r and the corresponding aspartate concentration. Make a plot of $r/$[aspartate] versus r.

C. What type of cooperativity is occurring? Which of the two limiting models discussed (MWC and KNF) is consistent with the data?

6-4. Consider a macromolecule that has two different conformations, M and M′. The two conformations bind a single ligand, L, per molecule, but the binding equilibrium is different for each conformation. This can be represented as

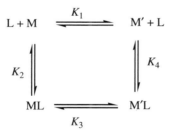

where the K_i are the equilibrium constants for the individual reactions. Calculate the binding isotherm, r, for this macromolecule in terms of the equilibrium constants and the concentration of unbound ligand. What type of cooperativity, if any, is displayed by this system? How many of the individual equilibrium constants can be determined from the binding isotherm? What relationship, if any, exists between the four constants?

Standard Free Energies and Enthalpies of Formation at 298 K, 1 Atmosphere, pH 7, and 0.25 M Ionic Strength

Substance	$\Delta G°$ (kJ/mol)	$\Delta H°$ (kJ/mol)
ATP	−2097.89	−2995.59
ADP	−1230.12	−2005.92
AMP	−360.29	−1016.88
Adenosine	529.96	−5.34
P_i	−1059.49	−1299.39
Glucose-6-phosphate	−1318.92	−2279.30
Glucose	−426.71	−1267.11
H_2O	−155.66	−286.65
NAD_{ox}	1059.11	−10.26
NAD_{red}	1120.09	−41.38
$NADP_{ox}$	1011.86	−6.57
$NADP_{red}$	1072.95	−33.28
Acetaldehyde	24.06	−213.97
Acetate	−247.82	−486.83
Alanine	−85.64	−557.67
Ammonia	82.94	−133.74
Ethanol	62.96	−290.76
Pyruvate	−350.78	−597.04
Formate	−311.04	−425.55
Sucrose	−667.85	−2208.90
Total CO_2	−547.10	−692.88
2-Propanol	140.90	−334.11
Acetone	84.89	−224.17
Glycerol	−171.35	−679.84
Lactose	−670.48	−2242.11
Maltose	−677.84	−2247.09
Succinate	−530.62	−908.68
Fumarate	−523.58	−776.57
Lactate	−313.70	−688.28

(continued)

Substance	$\Delta G°$ (kJ/mol)	$\Delta H°$ (kJ/mol)
Glycine	−176.08	−525.05
Urea	−39.73	−319.29
Ribulose	−328.28	−1027.12
Fructose	−426.32	−1264.31
Ribose	−331.13	−1038.10
Ribose 5-phosphate	−1219.22	−2042.40
Aspartate	−452.10	−945.46
Glutamate	−372.16	−982.77
Glutamine	−120.36	−809.11
Citrate	−966.23	−1513.66
Isocitrate	−959.58	—
cis-Aconitate	−802.12	—
Malate	−682.83	—
2-Oxoglutarate	−633.59	—
Oxalosuccinate	−979.06	—
Oxaloacetate	−714.99	—
Glycerol 3-phosphate	−1077.14	—
Fructose 6-phosphate	−1315.74	—
Glucose 1-phosphate	−1311.89	—
$CO_2(g)$	−394.36	−393.51
$O_2(g)$	0	0
$O_2(aq)$	16.40	−11.70
$H_2(g)$	81.53	−0.82
$H_2(aq)$	99.13	−5.02

This table is based on the conventions that $\Delta G° = \Delta H° = 0$ for the species H^+, adenosine, NAD^-, and $NADP^{3-}$ at zero ionic strength. These data are from R. A. Alberty, *Arch. Biochem. Biophys.* **353**, 116 (1998).

Standard Free Energy and Enthalpy Changes for Biochemical Reactions at 298 K, 1 Atmosphere, pH 7.0, pMg 3.0, and 0.25 M Ionic Strength

Reaction	$\Delta G°$ (kJ/mol)	$\Delta H°$ (kJ/mol)
$ATP + H_2O \rightleftharpoons ADP + P_i$	−32.48	−30.88
$ADP + H_2O \rightleftharpoons AMP + P_i$	−32.80	−28.86
$AMP + H_2O \rightleftharpoons Adenosine + P_i$	−13.55	−1.22
$2\ ADP \rightleftharpoons ATP + AMP$	−0.31	+2.02
$G6P + H_2O \rightleftharpoons Glu + P_i$	−11.61	−0.50
$ATP + Glu \rightleftharpoons ADP + G6P$	−20.87	−30.39

Data from R. A. Alberty, *Arch. Biochem. Biophys.* **353**, 116 (1998).

Structures of the Common Amino Acids at Neutral pH

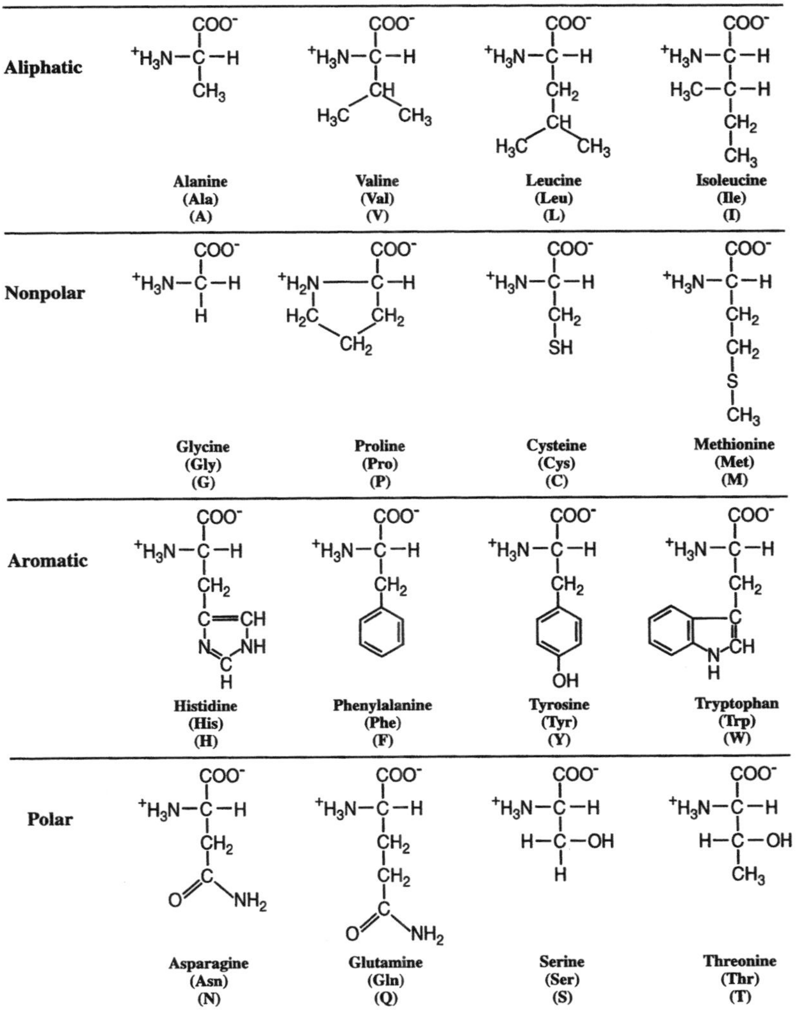

Aliphatic	$\overset{COO^-}{\underset{CH_3}{^+H_3N-C-H}}$	$\overset{COO^-}{\underset{\overset{CH}{H_3C\quad CH_3}}{^+H_3N-C-H}}$	$\overset{COO^-}{\underset{\overset{CH_2}{\underset{\overset{CH}{H_3C\quad CH_3}}{}}}{^+H_3N-C-H}}$	$\overset{COO^-}{\underset{\overset{CH_2}{\underset{CH_3}{}}}{^+H_3N-C-H \atop H_3C-C-H}}$
	Alanine (Ala) (A)	Valine (Val) (V)	Leucine (Leu) (L)	Isoleucine (Ile) (I)

Nonpolar	$\overset{COO^-}{\underset{H}{^+H_3N-C-H}}$	Proline ring structure	$\overset{COO^-}{\underset{\overset{CH_2}{SH}}{^+H_3N-C-H}}$	$\overset{COO^-}{\underset{\overset{CH_2}{\underset{\overset{CH_2}{\underset{\overset{S}{CH_3}}{}}}{}}}{^+H_3N-C-H}}$
	Glycine (Gly) (G)	Proline (Pro) (P)	Cysteine (Cys) (C)	Methionine (Met) (M)

Aromatic	Histidine structure	Phenylalanine structure	Tyrosine structure	Tryptophan structure
	Histidine (His) (H)	Phenylalanine (Phe) (F)	Tyrosine (Tyr) (Y)	Tryptophan (Trp) (W)

Polar	Asparagine structure	Glutamine structure	$\overset{COO^-}{\underset{H}{^+H_3N-C-H \atop H-C-OH}}$	$\overset{COO^-}{\underset{CH_3}{^+H_3N-C-H \atop H-C-OH}}$
	Asparagine (Asn) (N)	Glutamine (Gln) (Q)	Serine (Ser) (S)	Threonine (Thr) (T)

(continued)

157

Charged

COO⁻	COO⁻	COO⁻	COO⁻

$^+H_3N-C-H$ $^+H_3N-C-H$ $^+H_3N-C-H$ $^+H_3N-C-H$

Lysine: side chain $CH_2 - CH_2 - CH_2 - CH_2 - NH_3^+$

Arginine: side chain $CH_2 - CH_2 - CH_2 - N-H - C=NH_2^+ - NH_2$

Aspartate: side chain $CH_2 - C(=O)-O^-$

Glutamate: side chain $CH_2 - CH_2 - C(=O)-O^-$

Lysine	Arginine	Aspartate	Glutamate
(Lys)	(Arg)	(Asp)	(Glu)
(K)	(R)	(D)	(E)

Useful Constants and Conversion Factors

Gas Constant R	8.3144×10^7 erg K^{-1} mole^{-1}
	8.3144 joule K^{-1} mole^{-1}
	1.9872 calorie K^{-1} mole^{-1}
	0.082057 L atmosphere K^{-1} mole^{-1}
Boltzmann's constant k_B	1.3806×10^{-23} joule K^{-1} molecule^{-1}
Planck's constant h	6.6262×10^{-34} joule·second
Standard gravity g	9.8066 meter·second^{-2}
Electronic charge e	1.6022×10^{-19} coulomb
Faraday constant F	9.6485×10^4 coulomb mole^{-1}
1 calorie $= 4.184$ joule	
1 joule $= 10^7$ erg $= 1$ volt·coulomb	

INDEX